The Microcontroller Application Cookbook

Featuring the BASIC Stamp II[TM]

Matt Gilliland

Foreword by Bob DeMatteis, Inventor / Author

The Microcontroller Application Cookbook 2

Copyright © 2002 by Matthew Gilliland. All rights reserved.

Printed in the United States of America.

Except as permitted under the United States of America Copyright Act of 1976, no part of this publication may be reproduced or distributed in any form or by any means, or stored in a database or retrieval system, without the prior written permission of the author or publisher.

ISBN: 0-9720159-0-6

Note: Information contained in this book has been developed or obtained from sources believed to be reliable. However, the author cannot guarantee the accuracy or completeness of any information published herein and shall not be responsible for any errors, omissions, or damages arising out of the use or misuse of this information. This book is published with the understanding that neither the author nor the publisher is supplying engineering or other professional services. If such services are required, the assistance of an appropriate professional should be sought.

All trademarks are property of their respective owners and or company. BASIC Stamp and BASIC Stamp II are trademarks of Parallax, Inc.

The author can be reached at: matt@miconcookbook.com

This book is available at special quantity discounts to use as premiums and sales promotions, or for use in corporate or scholastic training programs. For more information, please contact Woodglen Press.

For updates to this book: www.miconcookbook.com

Published by Woodglen Press
www.woodglenpress.com
info@woodglenpress.com
(530) 432-3816

To Mom and Dad

Acknowledgements

Jen Jacobs, another great cover! Thank you!

Thanks to **Jameco** Electronics for providing many of the components necessary to create these circuits.

Thanks also to the staff at **Parallax**, for encouragement (again) and circuit suggestions.

My sincere appreciation to **Carol Valen** for taking the time to proofread my manuscript.

Thanks to my lovely wife, **Kimberly**, who said, "Sure, since you sold more than twelve copies of the first one, go ahead and write another" ☺

And finally, thanks to **all of you** who bought Volume 1 – you made Volume 2 possible!

Table of Contents

Foreword *by Bob DeMatteis* 7
    The American Inventor 9

Chapter 1: Introduction 11
    Second helpings 13

Chapter 2: Power Supplies 17
    Wall Transformer 19
    Power Transformer 29
    Battery 33
    Solar Power 39

Chapter 3: Power Control 43
    Supply Voltage Control 45
    Ground Circuit Control 61

Chapter 4: Input 71
    Switch 73
    Frequency Detection 87
    DTMF Detection 93
    Rotary Feedback 105
    Moisture Sensing 111
    Voltage Range Detection 115
    Pressure Measurement 123
    Linear Feedback 127
    Flow Detection 133
    Expanded Input 137
    External Event Counting 139
    Current Sensing 143
    AC Cycle Detection 147

Chapter 5: Output 151
    DC Motor 153
    Adjustable Flasher 167
    Audio Amplifier 171

	Watchdog Timer	175
	Stepper Motor	183
	D/A Conversion	187
	Voice Record and Playback	193
	Bar Graph Driver	195
Chapter 6:	**System Interfacing**	**207**
	RS-232	209
	RS-485 / RS-422	213
	Remote Control	219
Resources		**227**
Index		**235**

Foreword

Foreword

Foreword

The American Inventor

Creative, inventive people such as George Washington, Benjamin Franklin and John Hancock founded our country over 200 years ago. Little did they know that their creativity and inventive nature would be the foundation of a society that has evolved into the most prosperous country in the history of our planet.

Invention and creativity has always been the secret ingredient to America's prosperity and its prowess as a society and world power. Look everywhere and you can see the evidence. From inventions as simple as Eli Whitney's cotton gin, to Thomas Edison's light bulb, then Henry Ford's production line, and into the modern era with the numerous electronic inventions such as the cathode ray tube, the transistor, cell phone and even computer chips. All are American inventions, and are of immeasurable benefit to mankind.

You may not realize it, but you are also one of those inventive, creative humans who can make an impact on this world and the way we live. Oh, you may start out with simple daydreams and dabbling in some experiments. But keep in mind, that's exactly what Wozniak and Jobs did when they created Apple Computer, and what Bill Gates did when he created Microsoft. So did their predecessors, Whitney, Edison and Ford. No one knows where your creations will take you. However, one thing is certain, if you don't pursue your ideas and inventions, no one will benefit from them!

As an inventor, I know that ideas and inventions go through a metamorphosis. I know the significant amount of hours, weeks and months it may take to develop concepts into valuable, workable solutions...real time products. To see the

Foreword

end result is an exciting, rewarding accomplishment. When you create new applications for microcontrollers, you too have become an inventor. And you will experience tremendous joy with your accomplishments. You may even be able to patent your concepts, create new jobs and wealth and embark on a career—as an inventor—that you never before imagined possible. I know it can be done, because that is exactly what happened to me.

I also know that if you can shorten your development time by applying those concepts that are already known, have already been invented, you'd be wise to do so. At times, you simply do not have to "re-invent the wheel". When it comes to microcontrollers, Matt Gilliland is a prolific inventor too. Use his "recipes" and "cooking" experience to your advantage. Propel your inventive efforts forward and save hours, even days and months, by applying his tested, proven circuitry.

The circuits use simple, easy-to-find components that can be purchased from many electronics suppliers. Matt could make the circuitry more complicated, more sophisticated, but why? It's unnecessary and all accomplished inventors know the supreme beauty of applying simplistic methodologies to generate sophisticated results.

So, whether you are just beginning or want to learn how you can improve the performance of an existing project, you have become an inventor just like many other Americans from our historic past. Are you ready to make your mark in history?

Now...go out there and make it a better world!

Bob DeMatteis is the author of the book, *From Patent to Profit*, which teaches inventors how to turn their creative ideas into patents and profit at little or no cost. See "Resources" for details.

Chapter 1
Introduction

Chapter 1

Introduction

Second helpings...

The fact that you've picked up this book indicates that you have probably read its predecessor, *The Microcontroller Application Cookbook*.

If so, *welcome* to another installment of "real-world" microcontroller interfacing solutions! If not, might I *humbly* suggest adding the original *MAC* to your library. This *second volume* is designed to be a supplement to the first. Although some of the circuit titles may appear similar to those in Volume 1, the circuits (and code) are *different solutions* to common interfacing challenges.

The purpose of this "cookbook", as it was in the first volume, is to provide a hardware interfacing "databank" of circuits for microcontroller applications. My microcontroller of choice is the BASIC Stamp 2™, developed and manufactured by Parallax, Inc., in Rocklin, California.

However, you need not be a Stamp user to benefit from the circuits in this book. Every schematic (that works with the Stamp) can be adapted to just about any brand of microcontroller. Most hardware circuits will require no modification whatsoever.

The code samples are written in PBASIC – a simple to understand version of the popular BASIC programming language. If your microcontroller is something other than a Stamp, you'll of course need to modify the code.

Readers (of Volume 1) suggested some of the circuits in this book. As each of you have endeavored to create your own individual projects, countless hardware solutions have been

Chapter 1

devised and perhaps some of your suggestions have made it into this volume. Thanks to those of you who sent in circuit ideas.

I am aware that there are many different types of people reading this book – from educators and students, to robot builders and new product developers. No matter which category you fall into, we all have one thing in common - we love to tinker and create.

Just as in the original Cookbook, you won't find one complete program in the entire *Cookbook*. All I've done is give you a code snippet to demonstrate how the hardware circuit works. It's up to you to incorporate (or change completely) the code to fit your application. You are the "chef" and control which "ingredients" comprise your project. My intent is to "stir" your creative instincts with a myriad of circuit and code samples.

Those of you that have read the original *Cookbook* will notice that I've kept the same format here. That is to say, all schematics ("Figures") have an associated "Code" as well. For example, "Figure 3.4" works with "Code 3.4."

Each circuit has been constructed and tested numerous times with its associated Code. As mentioned above, the Code is just enough to show you how the circuit will operate, nothing more. Remember, I merely provide some of the ingredients - you're the chef!

These *Cookbooks* also assume you have some knowledge of basic electronics and the ability to breadboard or solder components into working projects.

Introduction

As an example, I assume you know what a resistor is, and how it works. If you don't have these skills, there are some excellent resources available to help you get up to speed.

Among these are:

Programming & Customizing the BASIC Stamp Microcomputer
Scott Edwards, TAB Books, ISBN #0-07-913684-2.

> Not only are there some fascinating (Stamp based) projects in this book, but the first half is an excellent beginning text on how to assemble and work with electronic circuits.

Microcontroller Projects with BASIC Stamps.
Al Williams, R&D Books, ISBN #1578201012.

> This is a great resource for learning about the Stamp. Many fascinating projects are included, as well as a more in depth description of the internal workings of the Stamp.

Stampworks
Jon Williams, Parallax, Inc. available at www.parallaxinc.com
ISBN #1928982077

> Thirty-one different experiments ranging from a simple blinking LED to a great introduction on serial communications within a microcontroller circuit.

www.parallaxinc.com
Parallax, Inc.
> More resources than I can mention here. This is the place to go if you're a "Stampaholic." They also

Chapter 1

maintain a "discussion list" that has a solid following of Stamp experts and novices alike.

www.stampsinclass.com
Parallax, Inc.

Another official Parallax website, but with more of an educational focus.

Some great learning materials here, available for free download. Among these are:

"What's a Microcontroller?"
"Analog to Digital"
"Robotics!"
"Industrial Control"
"Environmental"

Each of the above books is available in printed form for a nominal fee, or can be downloaded in .pdf format for free. They comprise a series of experiments or projects demonstrating how to use the Stamp in different types of applications. New resources are being added all the time - be sure to check back often.

It is my hope that these *Cookbooks* will provide a ready reference of simplified circuits that will expedite and enhance your ideas and projects.

Please drop me an email at: matt@miconcookbook.com, with any questions, comments, or suggestions for future volumes.

Happy Cookin'!

Matt

Chapter 2
Power Supplies

Chapter 2

Power Supplies

Wall Transformer

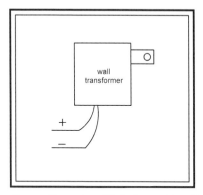

Wall transformers come in two basic types: *AC to AC*, and *AC to DC*.

The AC/AC version is simply a "step-down" transformer. This means that the typical 120 VAC household "current" input is "stepped-down" to a lower voltage – however the voltage is still "alternating." Figure 2.1 shows the output of a typical 9-volt AC/AC device.

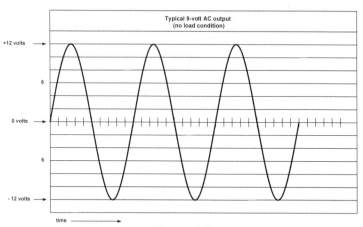

Figure 2.1
The typical output of a 9-volt AC/AC wall transformer (as viewed on an oscilloscope)

Chapter 2

Notice that the voltage alternates from a positive to a negative voltage (in relation to the "zero-volt" reference). Note also that the output *peak voltage* is somewhat greater than 9 volts.

Digital electronic circuits operate on DC, therefore we need to "rectify" the flow of electrons – that is to say, that we need to cause electricity to flow only in one direction.

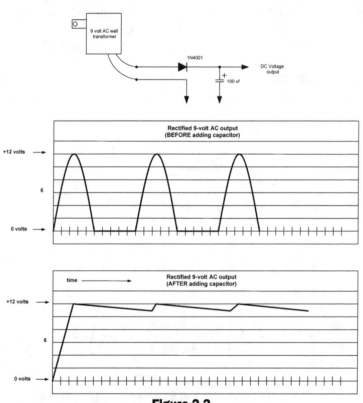

Figure 2.2
The typical output of a rectified 9-volt AC/AC wall transformer

Power Supplies

As shown in Figure 2.2, a diode can be used to rectify the voltage from AC to a "pulsed" direct current. The diode prevents the "negative" voltage from going through, resulting in a "pulsed DC" output.

Sensitive microcontroller circuits will not operate properly without a stable power source. Adding a relatively large electrolytic capacitor (as shown in Figure 2.2) smoothes out the pulses into a constant DC voltage.

The fact that there may be a slight "saw tooth" look to the wave shown in Figure 2.2 is not significant in most applications because, in most cases, we'll be regulating the voltage down to +5 volts anyway. We'll explore this in more detail later.

Figure 2.3 shows the same wall transformer, but the circuit employs a "full wave bridge" rectifier assembly. A full wave bridge is simply four rectifier diodes pre-connected within a single component package.

Chapter 2

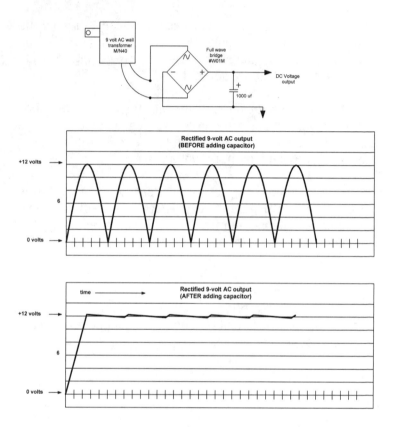

Figure 2.3
The output of a 9-volt full-wave bridge rectifier

You can create your own full wave bridge out of four independent diodes, as shown in Figure 2.4. Be sure to properly orient the diodes.

Power Supplies

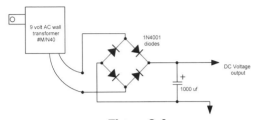

Figure 2.4
Building your own full-wave bridge rectifier from individual rectifier diodes

The schematic symbol of a diode is shown in Figure 2.5. The band on a "rectifier" diode is the cathode indicator.

Figure 2.5
The band indicates the cathode end of the diode

Line voltage fluctuations will cause the output of these circuits to go up and down as well. Since most electronic circuits need to operate from a stable power source, we need to add a voltage regulator as shown in Figure 2.6.

Chapter 2

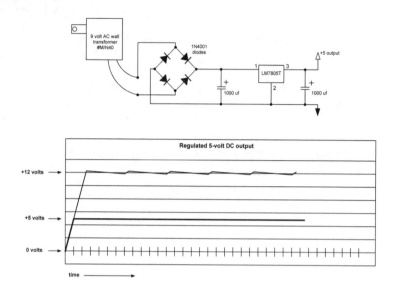

Figure 2.6
The LM7805T voltage regulator provides a stable voltage output, regardless of the input voltage conditions

The LM7805T is a three-pinned integrated circuit that is designed to control output voltage under changing input conditions. As you can see, the fluctuating voltage (around +12 volts) has no effect on the output of the LM7805T – its output is a steady +5 volts DC.

The LM7805T is capable of delivering up to 1 amp of output current. If your project uses less than 100 milliamps, you can save money and space by using the 78L05 regulator. Its function is identical to the LM7805T, but is physically smaller and delivers less current.

Power Supplies

The circuit in Figure 2.6 has limited output current due to the rating of the transformer. In this case, your circuit should draw no more than about 450 milliamps.

If you need more current than this, use the circuit in Figure 2.7. Its operation is essentially the same as 2.6 but uses components capable of delivering up to 1.5 amps of regulated +5 VDC output.

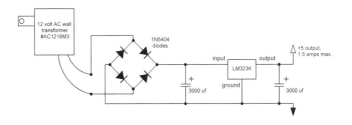

Figure 2.7
By using higher capacity components, the circuit can deliver up to 1.5 amps of current at a +5 VDC regulated output

Be sure to use an appropriate heat sink assembly and compound on the regulator; especially, when your power supply circuits must deliver high amounts of current.

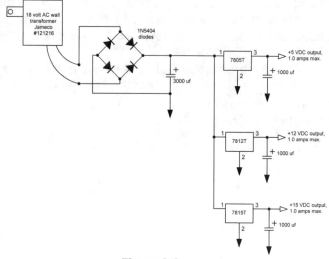

Figure 2.8
Multiple regulated voltage outputs, from a single wall transformer

If you need more than one voltage output, use the circuit in Figure 2.8. Although each regulator can deliver 1 amp of current, the overall amperage through all regulators should not exceed an aggregate total of 1 amp (the limit of the transformer).

An "AC to DC" wall transformer is shown in Figure 2.9. This type of device has built-in rectifier diodes, eliminating their need in the external circuitry.

Power Supplies

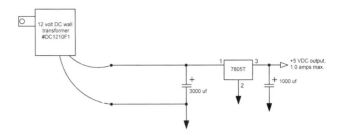

Figure 2.9
An AC to DC wall transformer eliminates the need for external rectifiers

Be sure to check the polarity of the wires on the transformer before connecting them to the regulator.

Chapter 2

Power Transformer

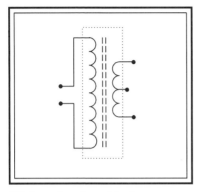

You can eliminate the wall transformer by creating your own "stand-alone" power supply, as shown in Figure 2.10. The circuit uses a "chassis mounted" power transformer.

There is no difference in operation from the prior "wall transformer" circuits. We've simply added a power cord and a separate chassis mounted transformer. Be sure to use proper grounding techniques (and caution!) when working with line voltages.

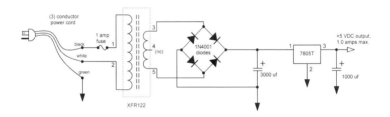

Figure 2.10
Power supply using a chassis mounted transformer and separate power cord

It's always good practice to include a fuse (or circuit breaker) when building and using power line operated devices. In-line fuse holders and fuses are "cheap insurance."

Figure 2.11 provides higher current capability, as well as several regulated voltage outputs.

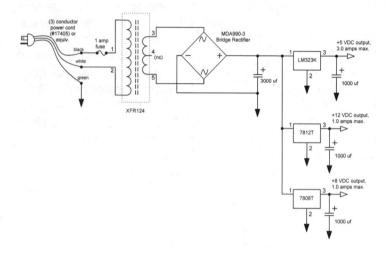

Figure 2.11
Higher amperage output, with multiple voltages, using a chassis mounted transformer and FWB Rectifier

Figure 2.12 uses an LM317T adjustable voltage regulator. Be careful when using this as a power source for any microcontroller circuit, as it is quite easy to adjust the voltage into the "expensive mistake" zone.

Power Supplies

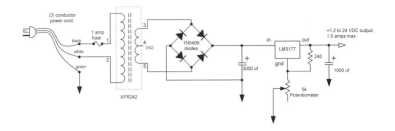

Figure 2.12
The LM317T provides a variable output voltage by adjusting the 5K potentiometer

Using the proper value resistors in the LM338T adjustment circuitry, results in a regulated 9 VDC output at up to 4 amps of current, as shown in Figure 2.13. If you wish to have a fully variable 3-amp power supply, simply replace the 1.5K resistor with a 5K potentiometer.

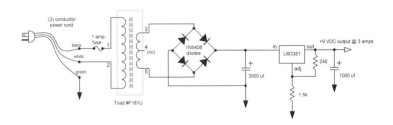

Figure 2.13
By using the proper value resistors, the LM338T provides a 9 VDC output @ 4 amps

If your project requires the use of a negative voltage (such as may be required in certain serial communication systems), you can use the circuit shown in Figure 2.14. The

Chapter 2

LM7912T operates in the same manner as the LM7812T, but from a negative voltage point of view.

In this case, the +/- 12-volt supplies could provide power for the serial communications portion of your circuit and the LM7805T delivers regulated +5 to the rest of the microcontroller system.

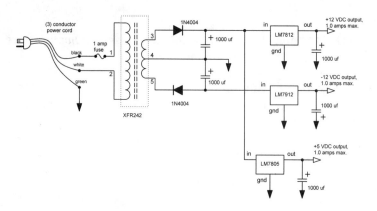

Figure 2.14
A power supply with three regulated outputs

Be sure to observe proper polarity on both the diodes and electrolytic capacitors (remember that ground is "positive" with respect to a negative voltage).

Power Supplies

Battery

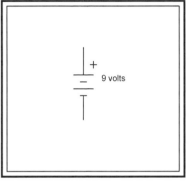

Batteries are great for portable device applications. Battery life, however, is completely dependent upon your circuit design. If your circuit "hogs" power, you'll need bigger batteries. Be careful in your component selection.

Technically speaking, a "battery" is a group of cells, each cell producing approximately 1.5 volts. If we connect these cells in series, the total voltage available is the aggregate sum of each cell's value, as shown in Figure 2.15.

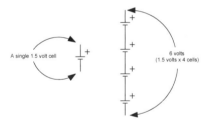

Figure 2.15
A single cell produces 1.5 volts. Four cells in series (battery) deliver 6.0 volts DC

If cells are connected in parallel, the voltage remains the same, but the amperage available increases as shown in Figure 2.16.

33

Chapter 2

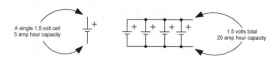

Figure 2.16
A single cell produces 1.5 volts. Four cells in parallel still deliver 1.5 volts, but the current capacity is quadrupled

As shown in Figure 2.17, the battery is rated at 9 volts, yet the schematic symbol only shows two "cells." Regardless of the number of cells in a battery, the schematic representation doesn't change. In other words, the symbol for both 9-volt and 12-volt batteries is the same. Typically the output voltage is shown next to the symbol, as depicted in Figure 2.17.

Figure 2.17
Schematic symbols for a single cell, and a battery (of any voltage)

Most microcontrollers (including the Stamp) are designed to operate on +5 volts. This supply voltage should be "regulated."

Typical voltage regulators require a minimum voltage input in order to function properly. The LM7805T needs a minimum of 7.5 to 8 volts input, otherwise the +5 volt regulated output is not guaranteed.

Power Supplies

Figure 2.18 uses a 9-volt battery. This could be a "standard" 9-volt (rectangular style) battery, or six 1.5-volt "D" cells connected in series. Obviously the "D" cells will have much more current capacity.

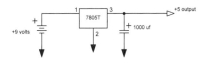

Figure 2.18
Regulating a 9-volt battery down to 5 volts

Figure 2.19 uses a zener diode to regulate the voltage to +5.1 volts. Due to the 1k resistor (in series with the load), current is somewhat limited.

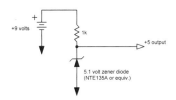

Figure 2.19
Using a 5.1-volt zener diode to regulate the voltage

Figure 2.20 uses five separate rectifier diodes to drop the voltage down to +5 volts. (Each diode "drops" about .8 volts times 5 diodes = 4 volts of drop).

This circuit may be cheap and work in a pinch, but its primary disadvantage is that it doesn't regulate at all - all it does is "drop" voltage. Therefore, as the battery voltage goes down, so too does the output.

Chapter 2

Figure 2.20
A "cheap" way to drop voltage

When using a regulator such as the LM7805T, even if no current is being drawn, the regulator itself will still consume power. In the case of the LM7805T, it's about 8 milliamps. This is not a big deal for "line voltage" operation, but on battery-powered projects, it may significantly reduce your expected current availability.

Careful circuit design can help to conserve power in battery-operated systems.

For example, the circuit shown in Figure 2.21 contains a switch connected as a sensor. The switch is a "normally-closed" momentary type.

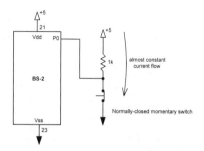

Figure 2.21
Power drain is relatively constant through the 1K resistor and normally-closed switch

Power Supplies

Code 2.21
    here:
    if in0=1 then there
    goto here

    there:
    debug "the door is closed"
    goto here

If this circuit were to be used as a "battery operated door closed sensor", and the door was usually open, then there would be a constant current flow down through the pull-up resistor and (closed) switch to ground. The current flow only stops when the door is closed (thus opening the switch).

One method to lower the amount of current drain, would be to increase the value of the resistor. A 10K resistor would decrease the current to one tenth (compared to the 1K shown in the circuit).

The manner in which your microcontroller program operates can also dramatically affect battery life.

As shown in Figure 2.22, the switch is a "normally open" type. Code 2.22 looks for the switch to be pressed (just like Code 2.21), but the detected input is a "low", therefore the code change: "if in0=0 then there."

This minor change in the program necessitated a change of switch type, but allowed the use of a 1K resistor as the pull-up device. Although there is still a small amount of current flowing through the 1K resistor into the Stamp, the switch is usually "open", and therefore minimizes current flow.

Chapter 2

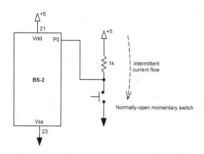

Figure 2.22
Power drain is intermittent through the 1K resistor and normally-open switch

Code 2.22
```
here:
if in0=0 then there
goto here

there:
debug "the door is closed"
goto here
```

Power management in battery-operated systems is critical for reliable circuit operation. It begins with good circuit design and the selection of low power consuming devices. Combine these elements with Code that takes advantage of the hardware, and your batteries will last much longer.

See Chapter 3 for additional circuits on power control and monitoring systems.

Power Supplies

Solar Power

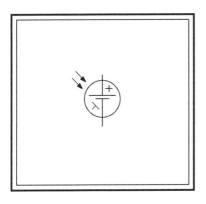

Unlike battery cells which produce approximately 1.5 volts each, a single solar cell only produces between .4 and .5 volts, and that's only when it's in direct sunlight.

Although this seems awfully restrictive, there are ways to overcome these hurdles, as we'll explore here.

The simplest way to get the needed voltage (typically 7 to 8 volts) is to use the "brute-force" method, as shown in Figure 2.23.

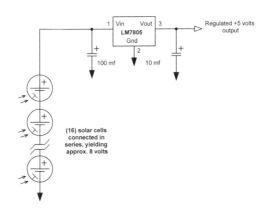

Figure 2.23
"Brute-force" method to get enough "solar cell voltage" for microcontroller applications

Chapter 2

Figure 2.24 utilizes a two-chip converter/regulator set from Maxim. The MAX660 doubles the input voltage (3.0 volts from the six solar cells) to 6.0 volts. The second chip (MAX 667) is a "low dropout" voltage regulator.

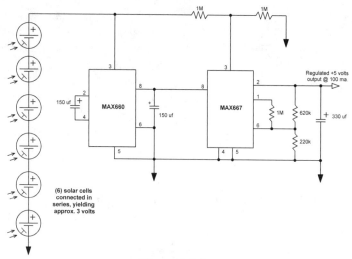

Figure 2.24
+5 volt, 100ma output from 3 volts of solar power

Together, these devices convert a low DC input voltage to a higher and regulated output voltage. In this example, we're going from six solar cells (yielding about 3.0 volts – depending upon lighting conditions), to a regulated +5 volts output. The output current of this circuit is limited to 100 milliamps. A significant advantage of this circuit is that is does not require any inductors.

Figure 2.25 uses two MAX660 connected in series. A single MAX660 can double the input voltage. By cascading two chips in series we can double the 1.5 volts to 3.0 and then again to 6.0 volts. The output is not regulated. See Figure

Power Supplies

2.26 for regulated +5 volt output, using the 78L05. You may have to add another solar cell to the circuit to get enough voltage output from the "doublers" for the 78L05 to regulate properly.

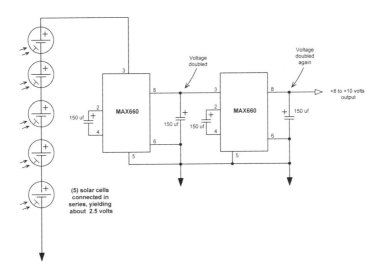

Figure 2.25
+6 volts (or more depending on sunlight conditions), 100 ma output from four solar cells

Figure 2.26 adds a low power version of the 7805 regulator series. Called the "78L05" ("L" for low power), the device is limited to a maximum of 100 milliamps output. It is available in a TO-92 type case (small "transistor sized" footprint).

Chapter 2

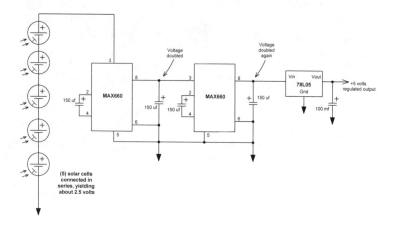

Figure 2.26
Regulated +5 volt, 100 ma output from five cells

Chapter 3
Power Control

Chapter 3

Power Control

Supply Voltage Control

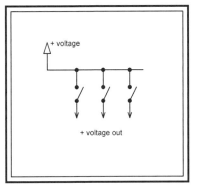

In some applications you may want to turn off power to selected portions of a circuit.

Supply voltage control can be as simple as a switch, as shown in Figure 3.1.

The power source is always constant to the microcontroller, but you can "shutdown" other portions of the circuit by flipping the switch, thus conserving power.

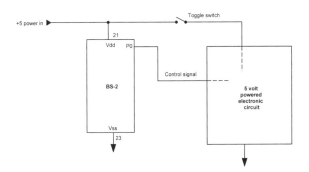

Figure 3.1
Simple power control of a portion of a project's circuit

Code 3.1
   "no code"

In certain applications, you may need to shutdown portions of your circuit under *program* control.

Chapter 3

For example, you may be building a sonar detection circuit in a mobile robot. If the robot isn't moving, then the onboard controller could turn off power to the sonar circuit, thus saving precious battery power.

Figure 3.2 uses a P-channel MOSFET (IRF9520) to source power to an isolated circuit.

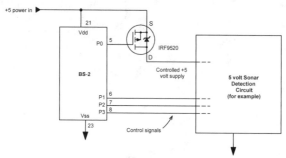

Figure 3.2
Microcontroller power control of a project's circuit

Code 3.2
```
     output 0
     here:
     low 0              'low turns ON power to circuit
     pause 2000
     high 0
     pause 2000         'high turns OFF power to circuit
     goto here
```

The P-channel device is turned "on" by driving the gate low to ground. For positive logic operation, you can add an inverter as shown in Figure 3.3. In this circuit, a "high" output from the I/O line turns power on.

Power Control

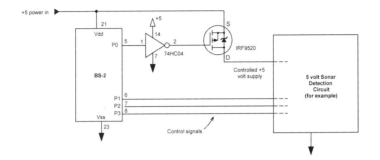

Figure 3.3
"Positive logic" controlled power of a circuit

Code 3.3

```
output 0
here:
high 0            'HIGH turns ON power to circuit
pause 2000
low 0             'LOW turns OFF power to circuit
pause 2000
goto here
```

The MOSFET acts as a high-efficiency switch, effectively controlling power to an entirely separate circuit. Since the voltage drop across the transistor is quite small (due to its low "on-state resistance"), the voltage delivered to the circuit when on, is very close to the supply voltage.

Figure 3.4 is the same concept with the addition of a latch and shift register. This circuit allows a 4-bit value (sent as serial data out of a single I/O pin) to control four separately powered circuits.

Chapter 3

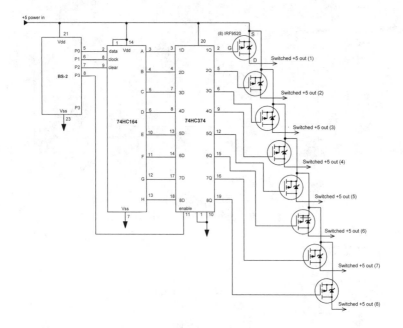

Figure 3.4
Eight separately controlled power sources using four I/O lines

Code 3.4a
```
    x var byte
    y var byte
    startover:
    y=1

    here:
    x=255-y              'first time through bits ="11111110"
    debug ? x            'which turns on circuit #1
    debug ? y
    shiftout 0,1,msbfirst,[x]
    pulsout 3,1
    pause 5000
```

Power Control

```
y=y*2          'send254,253,251,247,239,223,191,127
if x=127 then startover
goto here
```

Code 3.4b

```
y var word

here:
y=255                          'send out all 8 bits "high"
shiftout 0,1,msbfirst,[y]
pulsout 3,1                    'latch the data into the '374
debug ? y
pause 3000


y=247                          'bit pattern = 11110111
shiftout 0,1,msbfirst,[y]      'when the fourth bit = 0,
                               'circuit #4 is on

pulsout 3,1                    'latch the data into the '374
debug ? y
pause 3000

goto here
```

Code 3.4a simply sequences each of the power control transistors on and off. Code 3.4b turns on all eight power circuits, then after a delay of 3 seconds, turns off powered circuit #4 while leaving the other seven circuits on.

Logic on this circuit is inverted. This means that a "high" output from the micon's I/O line, will turn "off" the corresponding power circuit.

Chapter 3

In each of the prior circuits, there must be a common ground between both the "control" circuit and the "controlled" circuits. It is the positive voltage that is switched on and off via transistor (MOSFET) switches.

If you have circuits that do not (or cannot) have a common ground between them, you could use the circuit shown in Figure 3.5.

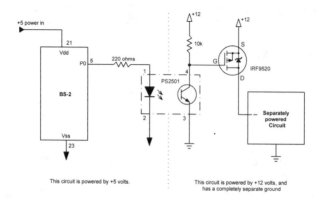

Figure 3.5
Controlling power with no common ground connection between power sources

Code 3.5
```
here:

    high 0              'turn ON the LED, which turns on power
    pause 1000
    low 0               'turn OFF the LED, which turns off power
    pause 1000

    goto here
```

Power Control

The optoisolator acts as a switch. When the LED inside the optoisolator is on, the MOSFET transistor is allowed to conduct (by pulling the gate "G" to ground), thus providing control via light to the separately powered circuit.

This type of power control can be quite useful in many types of battery-powered applications, such as mobile robotics.

Notice also that the "controlled" circuit can operate on a voltage other than +5 volts. In the sample circuit, we're using a +5 volt circuit to control a +12 volt circuit.

Figure 3.6 uses the MAX666 that has logic level shutdown capability. A high signal on pin 5 (of the 666) turns off the power output of the device. Although the current output is only 40 ma, the simplicity of this circuit is well suited for conserving power in battery operated applications.

Note that +12 volts is connected directly to the Stamp. This connection is on pin 24 (Vin), and takes advantage of the Stamp's on-board regulator. If your microcontroller of choice doesn't have a built in regulator, refer to the circuit in Figure 3.7.

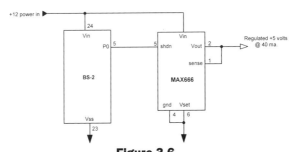

Figure 3.6
Fixed and regulated +5 volts from the MAX666

Chapter 3

Code 3.6
```
here:

high 0          'shut off the MAX666 regulator
pause 1000
low 0           'turn it back on
pause 1000

goto here
```

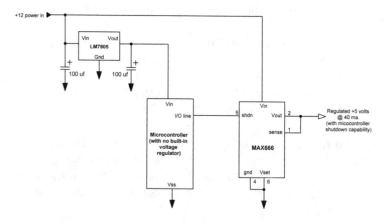

Figure 3.7
Similar circuit using a microcontroller that does not have an "on-board" regulator

Code 3.7
```
here:

high 0          'shut off the MAX666 regulator
pause 1000
low 0           'turn it back on
pause 1000

goto here
```

Power Control

Regulators like the MAX666 are widely used in pagers, remote data loggers, and many other similar types of devices. When shutdown or "off", the device only consumes about 12 micro amps.

Figure 3.8 uses the adjustable voltage output capability of the MAX666. By selecting the proper value of resistors (connected to pins 2 and 6), the device can output any regulated voltage between 1.3 and 15.0 volts. As shown below, the output is 9-volts. For another voltage use the formula shown in the figure.

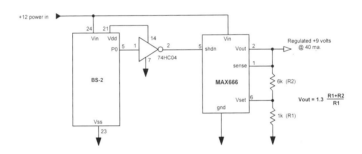

Figure 3.8
Adjustable voltage output capability, with micro-controlled shutdown feature

Code 3.8

```
here:

    high 0              'turn ON the MAX666 regulator
    pause 1000
    low 0               'turn it OFF
    pause 1000

    goto here
```

Chapter 3

The input voltage (to the MAX666) must be at least 1.0 volts more than your desired output voltage. This is called the "dropout" voltage (the difference between the input and output voltages).

For example, suppose your input voltage is 7 volts and you need an output of 6 volts. As long as the 7-volt input remains stable, regulated output is assured. However, if the 7 volt input drops at all, regulation is not guaranteed. You may still get 6 volts from the output, but it's not necessarily stable.

Figure 3.9 uses an LM317T adjustable voltage regulator. The output voltage is controlled by a "digital" potentiometer. See *MAC Volume 1* for more information on digital pots.

Under program control, the Stamp can vary the output voltage from 1.2 to +5 volts at up to 1.5 amps of current. For more current capacity use an LM338T, as shown in Figure 3.10.

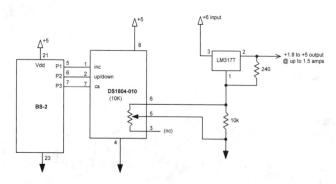

Figure 3.9
Programmable adjustable voltage output using the LM317T

Code 3.9

```
x var byte

low 3                'select DS1804-010
low 2                'set direction to counter-clockwise
for x=1 to 100       'reset the pot to "zero"
high 1
low 1
next

here:

high 2               'set direction to clockwise
for x=1 to 100       'ramp up output voltage over time
pause 20
high 1
low 1
next

pause 2000           'Maximum voltage reached

low 2                'set direction to clockwise
for x=1 to 100       'ramp down the output voltage
pause 20
high 1
low 1
next

pause 2000           'Minimum voltage reached

goto here
```

Chapter 3

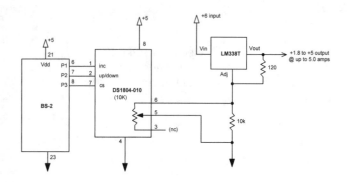

Figure 3.10
Five amp adjustable voltage power control using the LM338T

Code 3.10

```
x var byte
setpoint var byte

low 3              'select DS1804-010
low 2              'set direction to counter-clockwise
for x=1 to 100     'reset the pot to "zero"
high 1
low 1
next
setpoint=10

here:
high 2             'set direction to clockwise
for x=1 to setpoint 'ramp up voltage over time
pause 200
high 1
low 1
next

pause 2000         'setpoint voltage reached
low 2              'set direction to clockwise
for x=1 to 100     'ramp down the output voltage
```

Power Control

```
pause 20
high 1
low 1
next

pause 2000          'Minumum voltage reached
goto here
```

Figure 3.11 senses the amount of current that is flowing through a circuit. If current exceeds a preset level, then the MOSFET is turned off, similar to a circuit breaker except that the "trip point" is set in software. See Code 3.11.

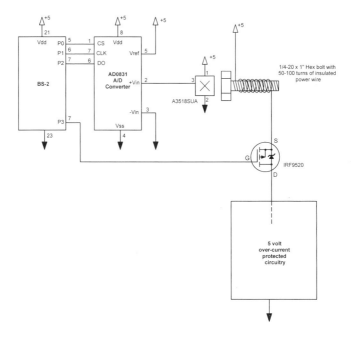

Figure 3.11
Detection and shut-down of an "over-current" condition

Chapter 3

Code 3.11

```
    x var byte
    y var byte
here:
    low 3                'turn power on to monitored circuit

    low 0                'enable the 0831
    pulsout 1,1          'send the first setup clock pulse
    x=0                  'set x to 0
    for y=1 to 8         'loop 8 times to get 8 data bits
    pulsout 1,1          'send a clock pulse
    x=x*2                'shift the bits left
    x=x+in2              'add x to the next incoming bit
    next
    high 0               'disable the 0831
    debug ? x            'display the result
    if x>140 then shutoff
    goto here            'go do it again

shutoff:
    high 3               'turn power off to monitored circuit
    pause 1000
    goto here
```

The "A3518SUA" is a linear hall-effect sensor, not a "switch" type. See MAC Volume 1 for more hall-effect information.

When current flows through the (insulated) power wire and P-channel MOSFET, a magnetic field is produced and concentrated by the hex bolt. By closely aligning the hall-effect sensor to the bolt, we can measure the current flowing through the wire by the intensity of the magnetic field. Be sure to use insulated wire when wrapping it around the bolt.

Power Control

Code 3.11 turns off the MOSFET if current exceeds a preset level, thus protecting the circuit from an "over-current" condition.

For instance, an over-current condition might occur when a drive motor on a robot stalls because of a collision event. Rather than continuing to "force" its way through, the circuit in Figure 3.11 will turn off the drive motor, thus saving precious battery power as well as not squishing the cat.

Chapter 3

Ground Circuit Control

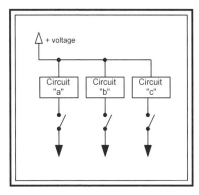

Figure 3.12 controls power by using an N-channel MOSFET. N-channel devices are optimized for circuits connected to the low side (ground) of a power supply.

Conversely, the P-channel MOSFET works well when connected to the high side of power, as evidenced in prior circuits. In general, N-channel devices (such as the IRF510 or IRF511) are less expensive and more readily available than their P-channel counterparts.

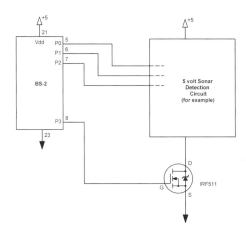

Figure 3.12
Simple ground circuit power control

Chapter 3

Code 3.12
```
here:
    toggle 3            'turn the power on and off
    pause 2000 'every two seconds
    goto here
```

If the "controlled circuitry" is a different voltage than what your microcontroller is operating on, you can use the circuit shown in Figure 3.13.

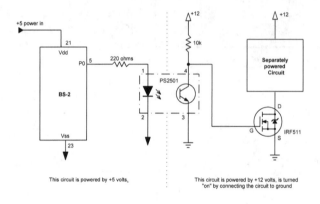

Figure 3.13
Optically isolated grounding control of a powered circuit

Code 3.13
```
here:
    high 0              'high turns OFF the power
    pause 3000
    low 0               'low turns ON the power
    goto here
```

Power Control

There are two significant advantages with this circuit, as compared with that of Figure 3.12.

The first is that the two individual circuits are completely isolated. Neither the power sources nor grounds of each circuit are connected in any way. This prevents any voltage spike or power glitch concerns, especially in controlled circuits that may have motors, solenoids, or other inductive loads.

Secondly, the "controlled voltage" doesn't need to be regulated (depending on the needs of the circuitry).

Good circuit design practice however, should be exercised whenever control schemes like this are implemented. As you can see in Figure 3.13, with no light (LED in the optoisolator is off) the IRF511 is pulled high by the 10K resistor. Pulling the MOSFET's gate high causes it to conduct. If the connection to the microcontroller were to fail (a broken wire at the optoisolator), then the circuit would be enabled. To prevent the possibility of a "run-away" condition we can add a simple transistor inverter as shown in Figure 3.14.

Now, if the optoisolator's input fails, the LED is off and so too, is the controlled circuitry. This is more of a "fail-safe" circuit design.

Chapter 3

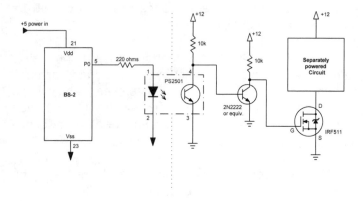

Figure 3.14
A more "fail-safe" circuit design

Code 3.14
```
here:
   high 0              'high turns ON the power
   pause 3000
   low 0               'low turns OFF the power
   goto here
```

Controlling multiple circuits by switching their ground connections can be implemented using the circuit in Figure 3.15.

The 74HC164 is an 8-bit serial shift register. Its output is latched by the 74HC374, which allows each grounding circuit to be individually controlled with a single bit within the serial string (sent by the microcontroller).

Code 3.15 sequences the outputs at two-second intervals.

Power Control

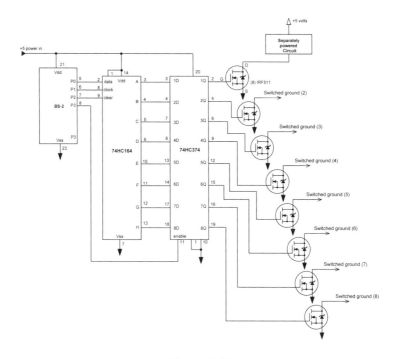

Figure 3.15
Controlling multiple ground power lines

Code 3.15

```
y var word

startover:
y=1
here:
shiftout 0,1,msbfirst,[y]     'shift the 8-bit value of y
pulsout 3,1                   'into the '164 and latch it
pause 2000
y=y*2                         'send 1,2,4,8,16,32,64,128
if y>128 then startover
goto here
```

Chapter 3

Figure 3.16 uses an optoisolator array that contains four independent emitter and detector pairs. Each power control circuit is switched "on" by a high output from the I/O pin. Inverters on the output side of the optoisolator won't actuate the circuit if a hardware fault on the input side occurs.

To prevent glitches from high current or inductive loads in the controlled power circuits, each side of the circuit (either side of the optoisolator) should run off its own +5 volts. Keep the grounds separate also. Instead of using the 74HC04, you could use a transistor inverter as shown in Figure 3.14.

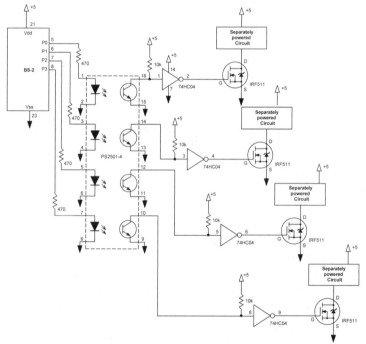

Figure 3.16
Optically isolated array controlling four ground circuits

Power Control

Code 3.16
```
dira=15
here:
outa=15         '1111 turns on all 4 circuits
pause 2000
outa=5          '0101 turns on 1st and 3rd circuits
pause 2000
goto here
```

Figure 3.17 adds a serial to parallel converter as well as a 74HC374 latch. The circuit allows the loading of a specific bit value without the "ripple' effect. The use of a couple optoisolator arrays provide for glitch-free circuit operation as well.

Code 3.17 steps through each corresponding "power enable" output.

Chapter 3

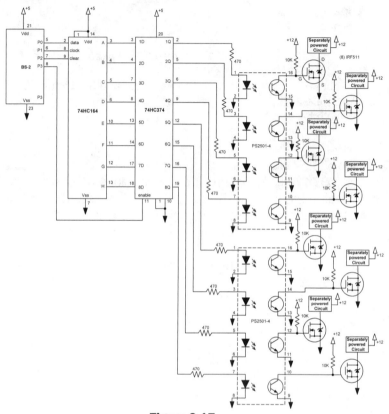

Figure 3.17
Optically isolated array controlling eight ground circuits, using only four I/O lines

Code 3.17
```
y var word

startover:
y=1
here:
```

Power Control

```
shiftout 0,1,msbfirst,[y]    'shift the 8-bit value of y
pulsout 3,1                  'into the '164 and latch it
pause 2000
y=y*2                        'send 1,2,4,8,16,32,64,128
if y>128 then startover
goto here
```

Figure 3.18 adds an inverter to each of the four power control circuits. The extra 7406 inverter must operate from +5 volts, thus the requirement for the separate LM7805 regulator shown in the upper right of the schematic.

You can avoid the need for the regulator circuit by using a simple 2N2222 transistor inverter, as shown in Figure 3.14.

Chapter 3

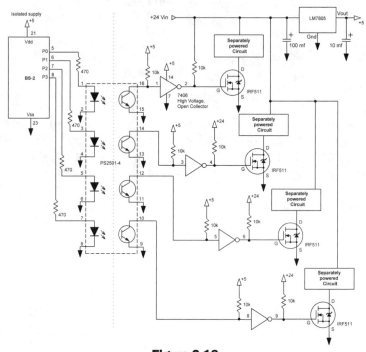

Figure 3.18
Optically isolated array controlling four ground circuits. The controlled circuits are powered by a separate and higher voltage than the control side of the optoisolator array

Code 3.18

```
high 0      'turn power ON to circuit #1
low 1       'turn power OFF to circuit #2
low 2       'turn power OFF to circuit #3
high 3      'turn power ON to circuit #4

stop
```

Chapter 4
Input

Chapter 4

Switch

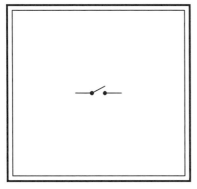

As we explored in MAC *Volume One*, switches come in a seemingly limitless array of styles and/or configurations. All accomplish essentially the same operation. However, switches "make" or "break" a connection between two portions of a circuit.

"Single-throw" switches make or break an electrical connection along a single conductive path. "DIP" switches contain multiple "single-throw" switches within the same device, as show in Figure 4.1. Each switching circuit is completely separate from the others, and can be *actuated* independently as well.

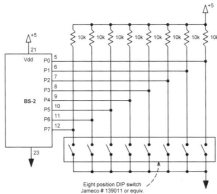

Figure 4.1
One byte "read" into a microcontroller using eight I/O lines

Chapter 4

Code 4.1
```
    x var byte
    y var byte

    here:
    x=inl           'get the 8 bit value
    y=255-x         'invert it

    debug ? x       'show me the values
    debug ? y
    debug cr
    pause 1000

    goto here
```

The circuit in Figure 4.1 "reads" the binary value of the eight position dipswitch array, and then saves it as a variable.

Unfortunately, this circuit uses a full 8-bit parallel port on the micon. For simple projects this may not be a problem, but running out of I/O lines before your project is completed is never fun.

By using a "data multiplexer", such as the 74HC151, we can reduce the required number of I/O lines down to four, as shown in Figure 4.2

Input

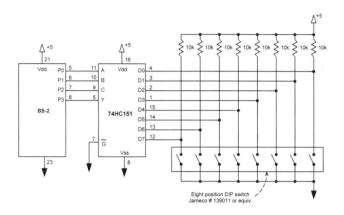

Figure 4.2
One byte "read" into a the Stamp using four I/O lines

Code 4.2

```
x var byte
y var byte
z var byte
dira=7              'set up P0-P2 as address outputs

here:
y=0                 'clear y before reading each input

for x = 0 to 7
outa=x              'select an input to read in sequence
z=in3
debug "input=",?x   'the address
debug "value=",?z   'the bit value of that address
debug cr

pause 200
y=y*2               'shift the value of y left
y=y+in3             'get value of the selected input and add
```

Chapter 4

```
next
debug "total = ",?y    'show the current value
debug cr
pause 3000
goto here
```

Figure 4.3 uses a 16 to 1 data multiplexer (the "74150") which allows a full 16-bit "word" to be read by the micon. The circuit only uses five I/O lines.

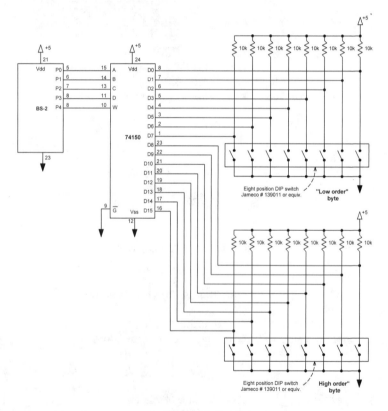

Figure 4.3
Reading a 16-bit word using only five I/O lines

Code 4.3

```
x var byte
y var word
z var bit
dira=15               'set up P0-P3 as address outputs

here:
y=0                   'clear y before reading each input

for x = 15 to 0       'step through each input
outa=x                'select an input to read
z=in4                 'get the value of that input
debug "input=",?x     'the address
debug "value= ",?z    'the bit value of that address
debug cr

y=y*2                 'shift the value of y left
y=y+in4               'get value of the selected input and add

next

debug "total = ",?y   'show the total
debug cr
pause 3000
debug cls

goto here
```

Thumbwheel (or push-wheel) switches provide a convenient method for users to input a value or parameter into a microcontroller. Rather than needing to understand binary (as required in the above circuits), the user only needs to "dial" or "punch-in" a decimal value. Check out Figure 4.4.

Chapter 4

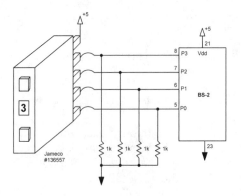

Figure 4.4
Push-wheel switch connections

Code 4.4
```
x var byte

here:

x=ina            'input 4 bits as one number
debug ? x

goto here
```

Be sure to include pull-down resistors in your circuit as shown in the schematic. The output of the push-switch relies on a pull-up (to +voltage) for proper circuit operation. When appropriate, each of the four output pins is connected to ground (by contact operation within the switch). A binary value is present on these four pins, which act as inputs to the micon.

Input

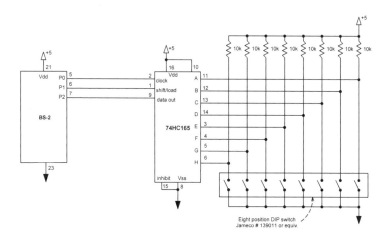

Figure 4.5
One byte input using three I/O lines

Code 4.5

```
x var byte
y var byte


here:
low 1             'read the 8-bit value into the '165
high 1

x=0               'clear x
for y=0 to 7      'cycle 8 times for the full byte
x=x*2             'shift bits left
x=x+in2           'add bits
high 0            'clock pulse
low 0
next

x=255-x           'invert the data
debug ? x         'show me
```

Chapter 4

```
pause 300
goto here
```

Figure 4.5 reduces the number of I/O lines to three in order to load an eight-bit value into the micon. This circuit uses the 74HC165 parallel to serial converter and an 8-position dipswitch, although you can use individual switches as well.

In fact, if you do use separate switches, you could monitor input conditions for multiple *locations*. For example, each switch could be connected to a different window in your home. Using only three I/O lines, the microcontroller could monitor the open/closed status of each individual window.

This circuit is essentially operating as a parallel to serial converter.

Figure 4.6 illustrates how to connect two push-switches to an 8 bit parallel to serial converter (74HC165). Only three I/O lines are required.

Input

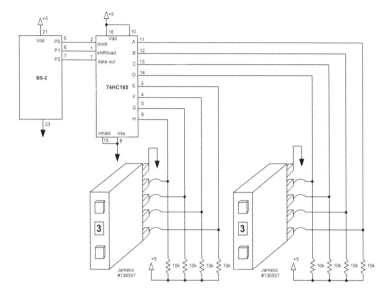

Figure 4.6
Two push-switches for "0 to 99" input

Code 4.6

```
x var byte
y var byte
z var byte
lower var byte
higher var byte

here:
low 1            'read the 8-bit value into the '165
high 1

x=0              'clear x
for y=0 to 7     'cycle 8 times for the full byte
```

Chapter 4

```
x=x*2               'shift bits left
x=x+in2             'add bits
high 0              'clock pulse
low 0
next

debug ? x           'show me the data before it's parsed

z= x<<4             'shift bit left 4 places
lower=z>>4          'shift bits right 4 places
                    'this clears out the higher digit so
                    'that it's not included in the lower
                    'digits value

debug ? lower       'lower order digit value

higher=x>>4         'shift bits right 4 places
                    'this moves the higher 8 bits to
                    'the lower 4 bit location for
                    'proper valuation

debug ? higher      'higher order digit value
debug cr

pause 1000
goto here
```

Figure 4.7 uses two 74HC165 serial to parallel devices. This allows for the input of a 16-bit word using only five I/O lines. The switches don't necessarily have to be "dips" – they could be just about any kind of device that outputs a "0" or a "1".

Input

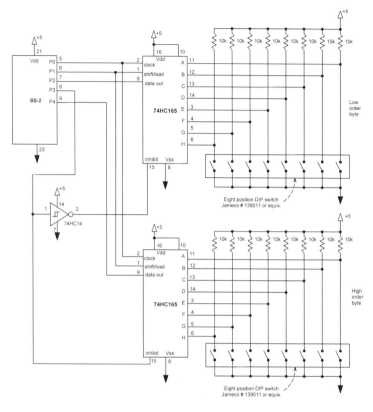

Figure 4.7
A 16-bit word input using five I/O lines

Code 4.7
```
x var byte          'low order byte
y var byte          'high order byte
z var byte
lower var byte
higher var byte

here:
```

```
high 3              'select the low order 74HC165
low 1               'read the 8-bit value into the '165
high 1

x=0                 'clear x
for z=0 to 7        'cycle 8 times for the full byte
x=x*2               'shift bits left
x=x+in2             'add bits
high 0              'clock pulse
low 0
next

x=255-x             'invert the data
debug ? x           'show me the low order byte


low 3               'select the high order 74HC165
low 1               'read the 8-bit value into the '165
high 1

y=0                 'clear x
for z=0 to 7        'cycle 8 times for the full byte
y=y*2               'shift bits left
y=y+in4             'add bits
high 0              'clock pulse
low 0
next

y=255-y             'invert the data
debug ? y           'show me the high order byte
debug cr


pause 1000
goto here
```

Using five I/O lines, the circuit in Figure 4.7 will input a full "word" (16 bits of parallel data) into the microcontroller.

Input

Also, you need not use dipswitch arrays for the input. You could use individual momentary switches located at different locations around the perimeter of a mobile robot. Using this circuit you could monitor 16 "collision" points, using only those five I/O lines.

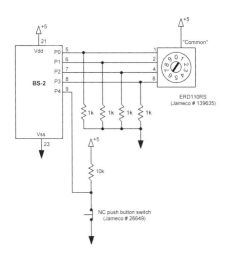

Figure 4.8
A rotary decimal input switch with "load" feature

Code 4.8
```
    x var byte
    y var word
    x=0

    here:
    if in4=1 then there    'loop until switch is pushed
    y=x * 50               'rotary switch can be set anytime
    debug ? y              'but value of y won't change until
                           'switch is pushed
```

Chapter 4

```
        goto here
        there:               'grab value of rotary switch
        x=ina
        goto here
```

If your application requires the input of a decimal value (0-9), you can use the circuit shown in Figure 4.8. A binary value of up to "1001" (decimal "9") can be selected by the user simply turning a dial – no need for binary number knowledge.

The normally-closed switch is used to alert the microcontroller when to read the value of the switch. Using this type of arrangement eliminates false values being read into the micon.

In other words, if your program was set up to read bits P0-P3 (the four bit value from the switch) each time through, then your project may experience erratic behavior if you were to go from "4" to say "8." By rotating the switch through positions 5, 6, and 7, your program may read this data "on the fly."

By only reading the four-bit value when the push-switch is actuated, you can turn the rotary switch to whichever value you desire and then "load" the value into the microcontroller.

Frequency Detection

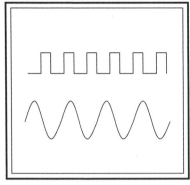

Frequency is a measure of the number of "cycles" of a repeating waveform within a given time period.

In most cases, frequency is expressed in "cycles per second", or more simply, "hertz." The average human ear can detect sound waves anywhere from 40 to 15,000 cycles per second. This is what we call "sound."

Not all "sound" is audible to the human ear. For example, *ultrasonic* frequencies are beyond our range of hearing, but some animals can hear frequencies well beyond humans. A "sound" in the range of 40,000 or 50,000 hertz is above the range of most humans, but dogs and other animals may be able to detect it over great distances.

The circuit in Figure 4.9 will measure the length of the low portion of a cycle.

Depending on the type of speaker used, you may be able to increase the frequency into the ultrasonic range. You may not be able to hear it, but the microcontroller can measure it (as demonstrated in Code 4.9).

Chapter 4

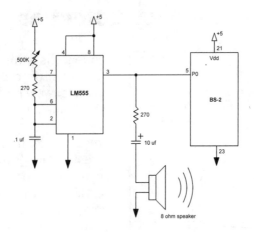

Figure 4.9
Simple frequency measurement

Code 4.9

```
x var word

here:
pulsin 0,1,x     'get the pulse width in 2 microsecond inc.
x=x*2            'double it
debug ? x        'show the width of the pulse in microseconds

goto here
```

Figure 4.10 adds one more device, and gives control of the frequency (produced by the LM555 circuit) to the micon through the use of a "solid state potentiometer." See Volume One for more circuits containing solid-state potentiometers.

Input

In this manner, you can set the output frequency of the 555 timer with code in the microcontroller, and then measure it to confirm the output.

The Stamp is not "tied up" to produce a series of pulses. The 555 timer handles that task. The frequency of the pulse stream can be altered at any time simply by toggling the solid-state pot.

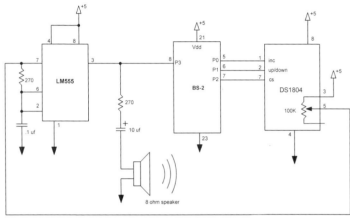

Figure 4.10
Control of output frequency with measurement and audio output capability

Code 4.10

```
x var byte
y var word

low 2              'select DS1804-010
low 1              'set direction to counter-clockwise
for x=1 to 100     'reset the pot to "zero"
high 0
low 0
```

next

here:
pause 2000

```
high 1           'set direction to clockwise
for x=1 to 100   'ramp up output voltage over time
pause 75
high 0
low 0
pulsin 3,1,y     'the width of the pulse in two
debug ? y        'microsecond intervals

next

pause 2000       'Maximum voltage reached

low 1            'set direction to clockwise
for x=1 to 100   'ramp down the output voltage
pause 75
high 0
low 0
pulsin 3,1,y     'the width of the pulse in two
debug ? y        'microsecond intervals
next

pause 2000       'Minumum voltage reached

goto here
```

Figure 4.11 replaces the LM555 tone source with a microphone. The incoming frequency is amplified and then digitized (converted to a square wave) by the 74HC14 Schmitt-trigger.

Input

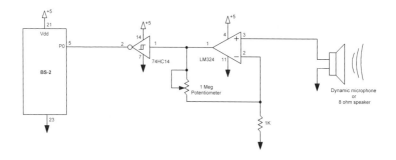

Figure 4.11
Measuring the frequency from an external audio source

Code 4.11
```
x var word

here:
pulsin 0,1,x    'get the pulse width in microsecond inc.
x=x*2           'double it
debug ? x       'show the width of the pulse in microseconds

goto here
```

To test this circuit, build a "stand-alone" 555-timer circuit that is connected to a small speaker. Move the speaker near the microphone. The sound waves emitted from the speaker travel through the air, are detected, amplified, and then measured by the Stamp.

Chapter 4

DTMF Detection

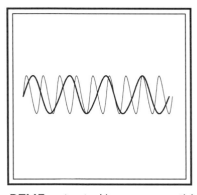

The Stamp 2 has a built-in command called "dtmfout." It is ideally suited for projects that require *generation* of DTMF tones.

Figure 4.12 along with the sample code show the basic connections and commands required to implement an audible DTMF output. You can use this simple *generation* circuit to build and test the following *detection* circuits. Of course, a standard touch-tone phone would work as well.

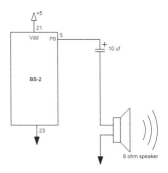

Figure 4.12
DTMF tone generation using the BASIC Stamp 2

Chapter 4

Code 4.12
```
x var word

here:
dtmfout 0,[5,5,5,1,2,1,2]     'dial 555-1212
pause 500

goto here
```

DTMF stands for "Dual-Tone-Multi-Frequency." This simply means that each time you press the "7" key on a touch-tone telephone, two distinct and separate tones are blended together resulting in a single audio output.

DTMF frequencies (in hertz) are as follows:

Key	Low Freq.	High Freq.
1	**697**	**1209**
2	697	1336
3	697	1477
4	**770**	1209
5	770	**1336**
6	770	1477
7	**852**	1209
8	852	1336
9	852	**1477**
0	**941**	1336
*	941	1209
#	941	1477
A	697	**1633**
B	770	1633
C	852	1633
D	941	1633

For example, by pressing the "7" key on a standard keypad, two independent frequencies are generated: 852 and 1209 Hz, respectively.

Input

You can see there are only eight distinct frequencies generated (underlined above), and all key "tones" are derived from a distinct combination of two of these.

Although possible to implement DTMF *detection* within a particular microcontroller, if your device doesn't have "built-in" capability, it's usually quicker (and certainly less "code intensive") to implement a hardware circuit solution such as that shown in Figure 4.13.

This circuit uses the M-8870-02 from Clare.

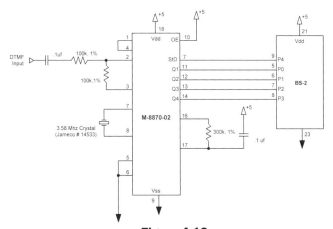

Figure 4.13
DTMF decoding using the M-8870-02 from Clare

Code 4.13
```
    x var word

    here:
    if in4=1 then get_value    'watch for valid data
    goto here
```

Chapter 4

```
get_value:
x=ina            'get the 4 bit data
debug ? x        'show me
goto here
```

Figure 4.14 adds a microphone and amplifier to the input of the decoder chip. This circuit, coupled with Code 4.14, decodes and displays the DTMF tones as they are produced. Use the circuit in Figure 4.12 or any touch-tone phone to generate the tone inputs.

Applications range from recording of all outgoing telephone numbers made from your phone to remote control devices.

Input

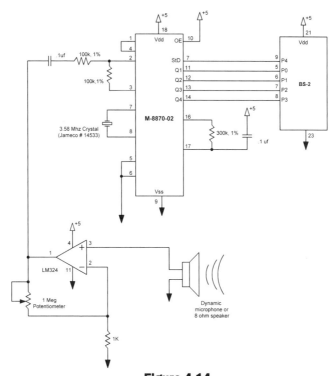

Figure 4.14
DTMF decoding with a microphone detection circuit

Code 4.14
```
x var word

here:
if in4=1 then get_value    'watch for valid data
goto here
get_value:
x=ina                      'get the data
debug ? x
goto here
```

Chapter 4

Figure 4.15 uses the CD22202 from Intersil. Its function is nearly identical to that of the M-8870-02 used in the prior circuit. The LM324 is configured as a non-inverting amplifier with gain adjustment via the 1 Meg pot connected between pins 1 and 2.

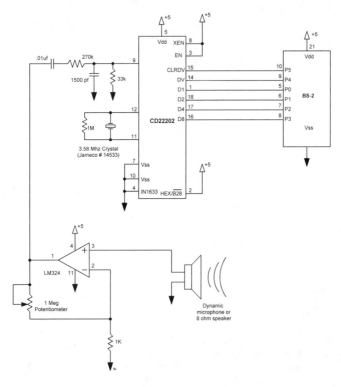

Figure 4.15
DTMF decoding using the CD22202 from Intersil

Code 4.15

```
x var word

here:
if in4=1 then get_value
goto here

get_value:
x=ina           'get hex value of tones
high 5          'resets DV signal
low 5           'with a handshake
debug ? x
goto here
```

Figure 4.16 uses a gaggle of individual tone decoders and a hoard of NOR gates to accomplish DTMF decoding. It's a significantly more complex circuit, but the components (decoder chips) may be more readily available.

Chapter 4

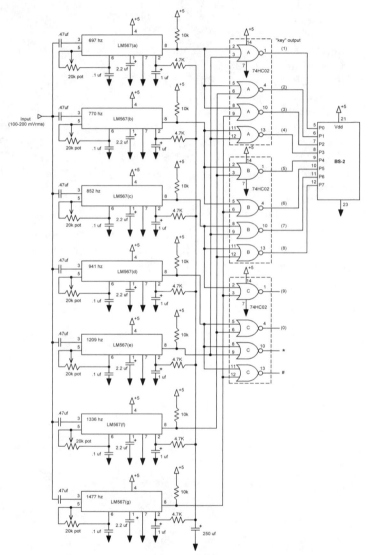

Figure 4.16
"Brute-force" method of tone decoding

Code 4.16

```
x var byte

here:
x = inl
if x=1 then show1        'determines where to go
if x=16 then show5
if x=128 then show8
goto here

show1:
debug "Touch tone key 1 detected"
goto here

show5:
debug "Touch tone key 5 detected"
goto here

show8:
debug "Touch tone key 8 detected"
goto here
```

As shown, each of the seven tone decoder circuit outputs are connected to an array of NOR gates. This results in single line outputs for each DTMF tone received. Figure 4.16 only shows (8) of these as inputs to the micon (outputs "1" through "8").

Although the microcontroller itself, shown in Figure 4.17, doesn't really decode DTMF tones (each LM567 circuit does that), it eliminates the need for the entire NOR gate array by letting the program determine which combinations of tone outputs are detected.

Chapter 4

Since keys "A – D" are not as widely used, Figures 4.16 and 4.17 only use seven tone decoder circuits. If you require the use of these additional keys, you'll have to add one additional tone decoder circuit, tuned for a frequency of 1633 Hz.

Input

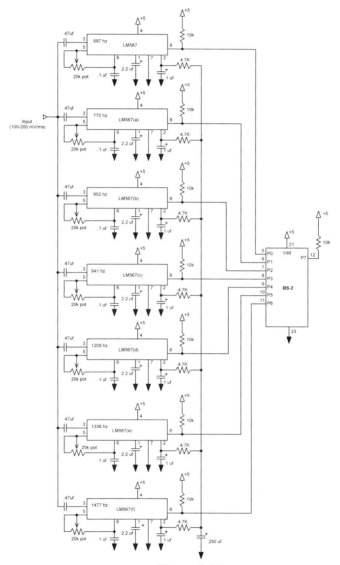

Figure 4.17
Alternative "software method" of tone decoding

Chapter 4

Code 4.17
```
x var byte
here:
x = inl
debug ? x
debug cr
if x = 221 then detect5    'looking for "11011101" =221
if x = 235 then detect7    'looking for "11101011" =235

goto here

detect5:
debug "5 key detected"
debug cr
goto here

detect7:
debug "7 key detected"
debug cr
goto here
```

Rotary Feedback

Rotary Feedback

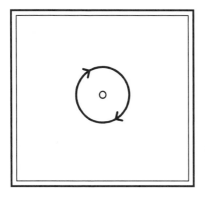

Rotary feedback to a microcontroller can be accomplished in many different ways.

If the device that you're building lends itself to having a "disc" attached to the rotating shaft, then you can use the circuit shown in Figure 4.18.

The disc is simply a round and flat piece of metal or plastic attached to the shaft or axel of the mechanical device. The disc has a series of holes punched or drilled through it to allow the passage of light.

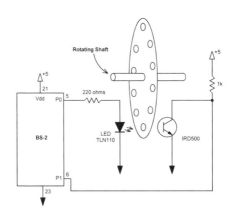

Figure 4.18
Simple method of detecting rotary motion

Chapter 4

Code 4.18a
```
    x var word

    high 0           'turn on LED
    here:
    if in1=1 then there
    goto here

    there:
    x=x+1            'increment x at each hole detection
    debug ? x

    recycle:
    if in1=1 then recycle
    goto here
```

Code 4.18b
```
    x var word

    high 0           'turn on LED
    here:
    pulsin 1,1,x     'measure the speed
    debug ? x        'lower number = faster

    goto here
```

Figure 4.18 uses commonly available LED's and phototransistors – almost any type will work.

You can increase the resolution, by adding more holes around the perimeter of the disc. Be sure to evenly space the holes to provide a more precise and consistent steam of pulses.

Figure 4.19 uses a "slotted optical detector." It performs the same function as the components used in Figure 4.18, but incorporates both the emitter and detector within one single device.

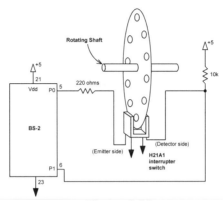

Figure 4.19
Using an "opto-interrupter" sensor

Code 4.19

```
x var word

high 0          'turn on LED
here:
if in1=1 then there
goto here

there:
x=x+1           'increment x at each hole detection
debug ? x

recycle:
if in1=1 then recycle
goto here
```

Chapter 4

Advantages are that the emitter and detector are "matched" together for optimized sensitivity. And since they're mechanically attached to the same plastic housing, alignment is assured.

The two primary disadvantages are that the device may be a little harder to find at your favorite electronics parts dealer, and that the disc itself must spin "true." That is to say, that the disc must be attached to the shaft so that there is no "wobble" during its rotation. This is due to the limited width of the slot (it's quite narrow).

The circuit in Figure 4.18 permits a looser tolerance and may be easier to implement. In either case, the Code is the same.

Figure 4.20 uses a hall-effect sensor to detect rotation. A small magnet is mounted on a rotating disk. Each time the disc rotates past the sensor a pulse is generated.

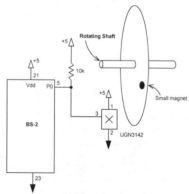

Figure 4.20
A hall-effect rotary sensor

108

Input

Code 4.20

```
x var word

here:
if in0=0 then there
goto here

there:
x=x+1      'increment x at each hole detection
debug ? x

recycle:
if in0=0 then recycle
goto here
```

To increase resolution, add more magnets onto the outer rim of the disc.

If your mechanical configuration does not lend itself to the attachment of a large disc to the shaft, you could try the circuit shown in Figure 4.21.

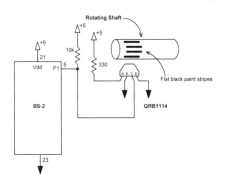

Figure 4.21
A reflective optical feedback sensor

Chapter 4

Code 4.21
```
x var word

here:
if in1=1 then there
goto here

there:
x=x+1        'increment x at each reflection
debug ? x

recycle:
if in1=1 then recycle
goto here
```

This type of sensor operates in the same manner as the preceding circuits, except that it is designed to detect the reflected light from its own internal emitter. Simply "draw" (with a black ink pen) some horizontal lines directly onto the rotating shaft. Alignment is critical, and ambient light can be a problem with this type of circuit. One way to help mitigate these problems is to place a shroud or covering over the entire assembly, thus blocking unwanted light.

If your shaft is a dark color, or made of a non-reflective material, you can add reflective strips of tape (like that used on bicycles or skateboards). Sometimes your circuit may even work better if you paint the shaft "flat black", and then add these reflective strips.

Moisture Sensing

Without water, life as we know it would not be possible.

However, excess water in the wrong places at the wrong time can be a problem.

A small water pipe leak under your house, gone undetected can cause significant rot problems.

Drying up the problem as soon as possible can significantly reduce moisture damage. Early detection is the key, and what better device to monitor "around the clock" than a micon?

Figure 4.22 is a very simple moisture detector. The sensor is simply two interlocking but not touching sets of "fingers." When water (or simply dampness) creates a bridge across two adjacent fingers, the MOSFET is turned "on." This pulls P0 to ground, thereby alerting the micon to a wet condition.

Chapter 4

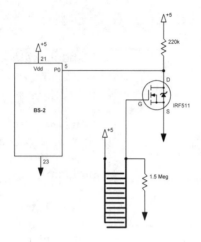

Figure 4.22
Simple moisture sensor

Code 4.22

```
    x var bit

    here:
    x = in0                     'get the value of P0
    if in0 = 0 then moisture    'decide what to do
    goto here

    moisture:
    debug "I sense moisture"    'so the basement is flooded
    debug cr
    pause 500
    debug cls
    goto here
```

Input

Figure 4.23 operates in a similar fashion, except that the circuit uses a P-channel MOSFET as the detection device.

The "finger array" can be constructed in numerous ways. On method is to simply use two bare wires close to, but not touching, each other. One other effective and reliable device is to simply etch your own PC board with that pattern.

Materials and instructions on how to etch your own PC boards are available through most electronic parts distributors, as well as your local Radio Shack store.

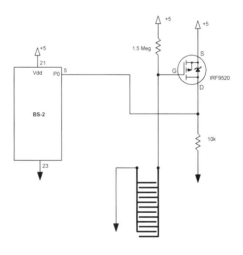

Figure 4.23
Moisture sensor using the P-channel IRF9520

Chapter 4

Code 4.23
```
x var bit

here:
x = in0
if in0 = 1 then moisture      'water here?
goto here

moisture:
debug "I sense moisture"      'so the plants don't need water
debug cr
pause 500
debug cls
goto here
```

Voltage Range Detection

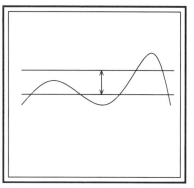

Microcontrollers are digital in nature and operation. If your project requires the micon to detect a certain analog voltage, you could include an analog to digital converter in the circuit.

An A/D converter measures a voltage and then encodes it into a digital value that the micon can understand. See Volume 1 for some easy to implement A/D conversion circuits.

In some cases however, you might not care about *what the value* of the analog voltage is, you may only need to know *when it is above or below* a certain set point.

Figure 4.24 uses an LM339 comparator. The comparator has two inputs. Pin 4 (the "minus" side input) is connected to the wiper arm of a standard potentiometer. Whatever the wiper arm voltage is, becomes the "set point." Now, if and when the input voltage on pin 5 exceeds the set point, the output of the LM339 goes "high."

So, for example, let's say that you have a water level sensor that has an analog output ranging from 0 to +5 volts. Assume that your project doesn't need to know exactly what the water level is. It just needs to know when the water drops below a certain level, such as, an automatic tank filling system.

Chapter 4

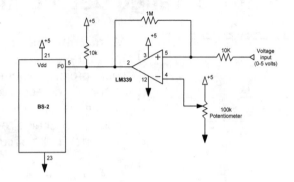

Figure 4.24
Voltage "threshold" detection

Code 4.24

```
x var bit

here:
x = in0                         'watch for overvoltage
if in0 = 1 then high_voltage
goto here

high_voltage:
debug "high voltage alert"
debug cr
pause 500
debug cls
goto here
```

Input

Instead of using an A/D converter that takes up several I/O lines, and an associated program to interpret the data, Code 4.24 reads the value of P0. If P0 is high, then there's enough water in the tank. If P0 is low, then the micon would turn on a valve, filling the tank and causing P0 to go high again – at which time, the valve is turned off.

Easy circuit and "no-brainer" code combine for a practical solution.

Figure 4.5 adds another comparator to the circuit. The LM339 has four contained within the same package. There are two set points, each set by independent potentiometers.

The "upper" comparator operates as the "under-voltage" set point and the "lower" one (Pin 1 output on the LM339) is set as the "over-voltage" set point.

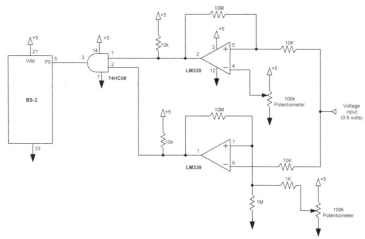

Figure 4.25
Detecting a pre-set voltage range

Chapter 4

```
here:
x = in0                    'check for voltage level
if in0 = 0 then out_of_range
goto here

out_of_range:
debug "The voltage is out of range"
debug cr
pause 500
debug cls
goto here
```

By adding the AND gate, both conditions (outputs from each of the comparators) must be "high" for a "high" input result on P0 of the micon.

The top comparator outputs a "high" signal as long as the voltage input does not go below its set point. The bottom comparator outputs a "high" as long as the voltage input does not exceed it's set point. Therefore, as long as the voltage input stays within a certain range (determined by the two potentiometers), the output of the AND gate is "high."

If either set point is breached, the corresponding comparator outputs a "low" resulting in a logic "1" at the P0 input of the micon.

Figure 4.26 adds greater flexibility to the above circuit.

Input

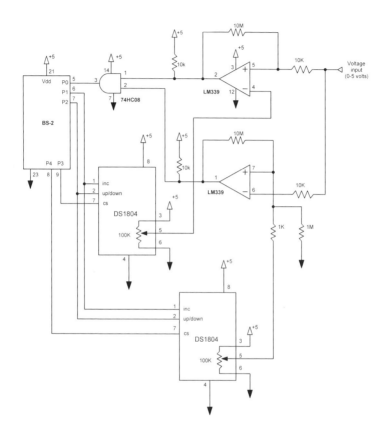

Figure 4.26
An "over/under" detector with micro-controlled set points

Code 4.26
x var byte
y var bit

low 3   'select the first DS1804-010
low 2   'set direction to counter-clockwise

Chapter 4

```
for x=1 to 100    'reset the first pot to "zero"
high 1
low 1
next
high 3

low 4             'select the second DS1804-010
low 2             'set direction to counter-clockwise
for x=1 to 100    'reset the second pot to "zero"
high 1
low 1
next
high 4

here:
low 3
high 2            'set direction to clockwise
for x=1 to 33     'set voltage to 1/3 V+ on first pot
high 1            'change '33' to between 1 and 100
low 1             'for 'under voltage' setpoint
next
high 3

low 4
high 2            'set direction to clockwise
for x=1 to 66     'set voltage to 2/3 V+ on first pot
high 1            'change '66' to between 1 and 100
low 1             'for 'over voltage' setpoint
next
high 4

check_range:
x = in0
if in0 = 0 then out_of_range
goto check_range

out_of_range:
debug "out of range alert"
debug cr
pause 100
```

```
pause 100
debug cls
goto check_range
```

As you can see, a solid-state device has replaced each manual potentiometer. Under program control, the set point conditions can be altered "on the fly", if necessary.

Chapter 4

Pressure Measurement

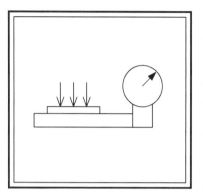

Measuring small amounts of pressure can be quite useful in many situations. For example, you may wish to have your micon monitor the pressure in the air pump of your aquarium tank.

No pressure would indicate a pump failure and a pressure-monitoring device could save countless (fish) lives.

Figure 4.27 uses an inexpensive pressure sensor available through Jameco. It can comfortably measure between zero and five PSI. The sensor's output is an analog voltage ranging from 0 to 4.5 volts.

The AD0831 is an 8 bit A/D converter and can easily convert this analog voltage to a digital value, as shown in Code 4.27.

Chapter 4

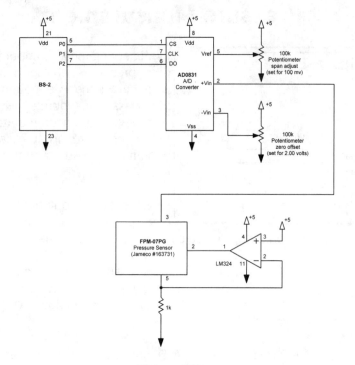

Figure 4.27
A simple pressure measurement circuit

Code 4.27

```
x var byte
y var byte
here:
low 0              'select the converter
pulsout 1,1        'send the first "setup clock pulse"
x = 0              'set x to zero
for y = 1 to 8     'loop 8 times to get the 8 data bits
pulsout 1,1        'send a clock pulse
x = x*2            'shift bits once to the left
x = x+in2          'add x to the incoming bit
```

```
next            'do it 8 times
high 0          'de-select the ad0831
debug ? x       'print the result
goto here
```

Alternatively, if you only need to know when the pressure has dropped below or gone above a particular set point, you could integrate a circuit such as those from Figures 4.24 or 4.25.

Chapter 4

Input

Linear Feedback

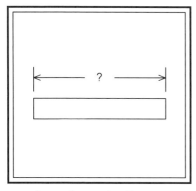

If you need to know when the pneumatic piston or hydraulic ram of a robotic arm is fully extended, you could use one of the following circuits.

Linear motion is sometimes difficult to achieve, but once accomplished, it may be even more important to monitor where the linear device is in relation to its "home" position.

Figure 4.28 depicts one method of tracking the actual location of an air cylinder's piston.

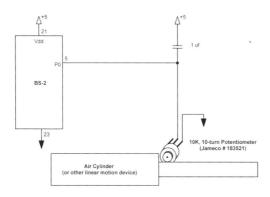

Figure 4.28
Rubber wheel attached to a potentiometer for linear feedback

Chapter 4

Code 4.28
```
    x var word
    here:
    high 0          'reset the capacitor to zero
    pause 1
    rctime 0,1,x    'get the RC time value
    debug ? x       'show me
    goto here
```

The circuit takes advantage of the Stamp's built in command "RCtime", which measures the length of time it takes to charge up a particular value capacitor through a particular value resistor. When that threshold is reached, the time displayed (via Code 4.28) is relative to the value of the potentiometer.

The pot should have a rubber wheel of some type that is permanently affixed to its shaft. If the extended travel of the linear rod is greater than the circumference of the wheel, you could either use a larger diameter wheel or use a multi-turn pot (such as that shown in Figure 4.28).

If your microcontroller does not have an "RCtime" type of command, you could use the circuit shown in Figure 4.29. In this case, the potentiometer's resistive value controls the frequency of the pulses being generated by an LM555 time running in "astable" (free-running) mode.

Code 4.29 measures the period between the end of one cycle and the beginning of another. As the value of the pot changes (relative to the movement of the linear motion device), the position of the shaft can be determined.

Input

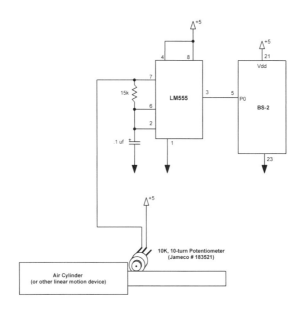

Figure 4.29
Using pulse-width modulation for linear movement detection

Code 4.29
```
x var word
here:
pulsin 0,1,x     'measure the pulse width determined by
debug ? x        'the rotation of the pot
goto here
```

Figure 4.30 uses a completely different method of determining the relative location of the linear shaft.

Rather than tracking the position of the shaft throughout its travel, the reflective optical sensors detect its minimum and maximum positions.

Chapter 4

Two sensors are used. One for "retracted" (the sensor on the bottom of the schematic) and one for "extended." The "indicators" on the shaft are either a strip of reflective tape, or in the case of an already reflective shaft, black ink.

One advantage of this circuit is that it's completely solid-state – there are no moving parts to wear out. Of course, if your project entails the use of air or hydraulic cylinders, the tape or ink markings may rub off.

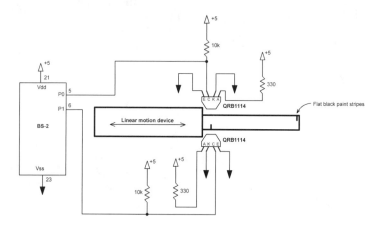

Figure 4.30
Reflective optical linear movement feedback

Code 4.30
```
here:
    if in0=1 then retracted        'looking for no reflection
    if in1=1 then extended
    debug "the device is in the middle somewhere"
    debug cr

    goto here
```

retracted:
debug "retracted"
debug cr
goto here

extended:
debug "extended"
debug cr
goto here

Figure 4.31 uses mechanical switches to accomplish the same results as Figure 4.30. Unfortunately, it's not solid-state, but there's no ink or stickers to rub off either.

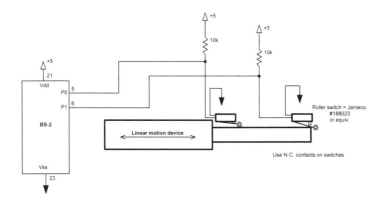

Figure 4.31
Mechanical switch limit detection

Code 4.31
```
here:
    if in0=0 then retracted    'looking for non-actuated switch
    if in1=1 then extended     'looking for actuated switch
```

```
    debug "the device is in the middle somewhere"
    debug cr

goto here

retracted:
debug "retracted"
debug cr
goto here

extended:
debug "extended"
debug cr
goto here
```

Flow Detection

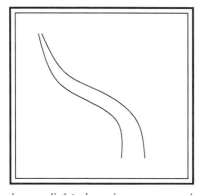

Determining when air or water is actually flowing through a pipe or tube can sometimes be difficult to determine by conventional means.

For example, suppose you have a water pipe that is pressurized to 5 PSI. When you open a valve there may be a slight drop in pressure, but if your pumping system is working properly, it should be hardly noticeable (unless you open the valve all the way). And of course, "hardly noticeable" means "more difficult to detect."

Just because there's a drop in pressure doesn't necessarily mean that there is an actual *consistent flow* of liquid.

Flowmeters are available, but the inexpensive ones usually consist of a dial or some other type of visual indicator. More expensive types have analog or even digital outputs that can be easily interfaced to a microcontroller using other circuits in these *Cookbooks*. But again, they tend to be on the expensive side.

Figure 4.32 shows a rather inexpensive device that can be obtained from numerous plumbing supply companies.

Chapter 4

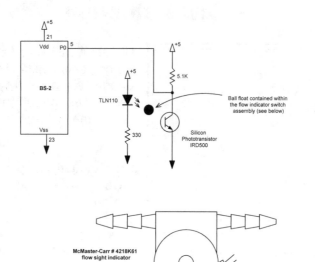

Figure 4.32
A water/air flow indicator circuit

Code 4.32

```
here:
if in0=0 then airflow          'ball not blocking light?
debug "there is no flow"
debug cr
goto here

airflow:
debug "air or water is flowing"   'ball has movement
debug cr
goto here
```

Whenever there is a flow of water or air, the little ball in the device moves back and forth, depending on the direction of flow. The more extreme the flow rate the further the ball moves.

By using a small linear array of LED's and phototransistors, you can get a relative idea of how much flow there is.

Chapter 4

Input

Expanded Input

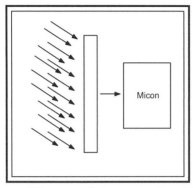

The *Cookbook* Volume 1 has several different circuits that show how to gain additional input sensing capabilities.

I have added another option here. By utilizing a single chip (the 74150 - 16 to 1 multiplexer), you can detect 16 different I/O conditions, utilizing only five I/O lines, as shown in Figure 4.33.

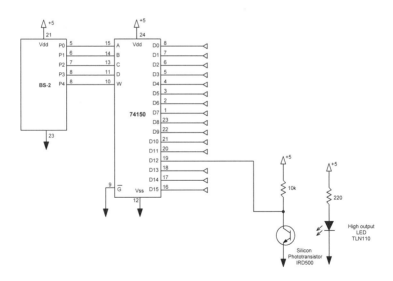

Figure 4.33
16 input lines using only five microcontroller I/O lines

Chapter 4

Code 4.33
```
y var bit
dira=15     'all four bit are used for address
here:
y=0
outa=12     'look specifically at D12 input
y=in4       'get the bit value

debug ? y
debug cr

goto here
```

As you see in the schematic, any number (and different types) of devices can be connected to the inputs of the 74150.

Circuits like this help to add "virtual" input capability to your micon circuitry, but they do require more programming skills to "keep it all straight."

Essentially, your program needs to set an "address" and then "read" the value of the input. Of course, the program needs to keep track of which input it's monitoring at any given instant and respond appropriately.

External Event Counting

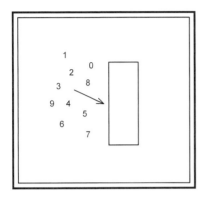

Yes, microcontrollers can easily count "events", such as high and low transitions on one of its I/O lines.

However, there is a concern that the micon must be continually monitoring that line, otherwise it may miss an "event."

Figure 4.34 ues an external counter to keep track of how many pulses have occurred over a given period of time. The counter chip (74HC193) is dedicated to incrementing its internal counters each time there's a transition on its "Up" input line.

In the example shown, the circuit is set up to count the number of times that an "electric eye" circuit is tripped. However, you can use any "digital event" as the input to the counter chip. For example, you could monitor the number of times that a doorbell "button" was pushed before somebody finally answers the door.

Chapter 4

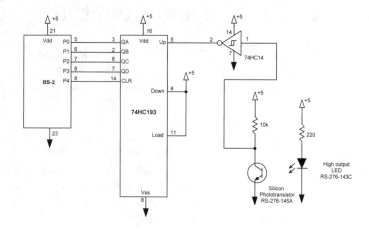

Figure 4.34
Counting external "optically interrupted" events

Code 4.34

```
y var byte

high 4          'reset counter to zero
low 4

here:
y=ina           'get value of P0-P3
debug ? y
debug cr

goto here
```

Figure 4.35 uses the "Down" input pin on the 74HC193. You can preset a binary value from which the counter can decrement.

Input

As an example, Code 4.35 loads "8" as a preset. For each input pulse, the counter is decremented. When the counter reaches 0000, "BO" goes low.

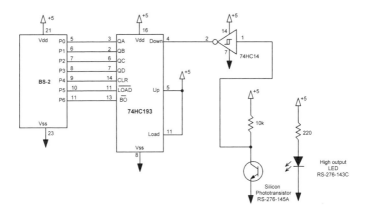

Figure 4.35
Counting "down" events

Code 4.35

```
    y var byte

    high 5          'deactivate load function
    dira=15         'set address bits as outputs

startover:
    high 4          'reset counter to zero
    low 4

    outa=8          'use any number 1-15
    low 5           'load the counter with that number
    high 5
```

```
here:
y=ina                'get value of P0-P3
debug ? y
debug cr
pause 3000           'verify that # was loaded

check_for_zero:
if in6 =0 then counter_at_zero
debug "checking for zero"
debug cls
goto check_for_zero

counter_at_zero:
debug "counter reached zero"
debug cr
pause 1000
goto startover
```

Input

Current Sensing

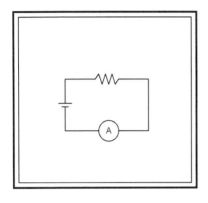

The flow of electrons through a wire is called "current." The basic unit of measurement for current is the "amp."

Knowing whether or not a current is flowing through a particular conductor can be a useful indicator in many types of applications.

For example, let's assume that your circuit is supposed to turn on an electric motor. However, when the microcontroller sets its I/O bit to "on", the motor doesn't work. By being able to detect whether or not current is flowing, the microcontroller could help to diagnose the problem. No current flow means that there is a power, fuse or conductor problem. However, if current is flowing but the motor isn't turning, then there may be a mechanical "jam."

Figure 4.36 demonstrates a very simple method of implementing a current flow detection circuit.

It operates on the basic principle that whenever electricity flows through a conductor, an electro-magnetic field is created. By wrapping the conductor around a nail, bolt or other piece of ferrous metal, we can concentrate the field enough to be able to detect it with a hall-effect switch. See MAC Volume 1 for more circuits based on hall-effect sensors.

Chapter 4

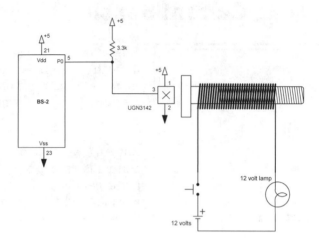

Figure 4.36
Hall-effect current flow detection

Code 4.36
```
    x var bit

    here:
    if in0=1 then there     'check for magnetic field
    goto here

    there:
    debug "there is current flowing"
    debug cr
    goto here
```

When the switch (to the light) is pressed, current flows creating a magnetic field that is detected by the sensor, which, in turn, produces a "low" output to the micon's I/O line. Code 4.36 merely "looks" at the I/O line when it's convenient for the program to do so.

Input

Figure 4.37 uses a similar detection scheme. But by utilizing a linear hall-effect device, provides an analog voltage output relative to the amount of current flowing through the motor. The analog voltage is converted to serial digital data for subsequent processing by the microcontroller.

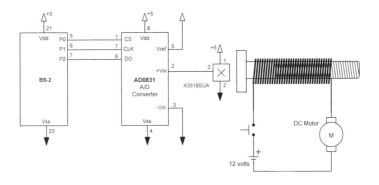

Figure 4.37
Hall-effect current flow measurement

Code 4.37

```
x var byte
y var byte
here:
low 0              'select the converter
pulsout 1,1        'send the first "setup clock pulse"
x = 0              'set x to zero
for y = 1 to 8     'loop 8 times to get the 8 data bits
pulsout 1,1        'send a clock pulse
x = x*2            'shift bits once to the left
x = x+in2          'add x to the incoming bit
next               'do it 8 times
high 0             'de-select the ad0831
debug ? x          'print the result
goto here
```

Chapter 4

AC Cycle Detection

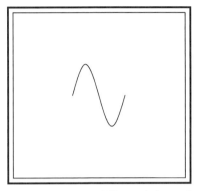

The detection of AC voltage can be accomplished several ways.

The safest method is to simply use a wall transformer as shown in various circuits in Volume One.

However, if your application requires the ability of knowing not only whether or not there's voltage, but also which point in the AC cycle you're at, you could use a circuit like that shown in Figure 4.38.

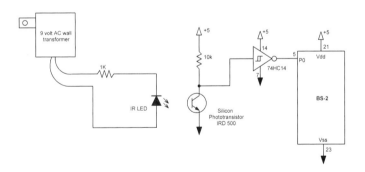

Figure 4.38
Simple AC cycle (zero cross) detection

Chapter 4

Code 4.38a
```
here:
if in0=0 then there     'LED is off right now
goto here

there:
debug "cycle beginning"
goto here
```

Code 4.38b
```
x var word

here:
pulsin 0,0,x
            'whichever command is inserted here starts
            'on the beginning of an AC cycle
debug ? x
goto here
```

The wall transformer must be an "AC to AC" type – not the other, more common "AC to DC" model. In many "DC" models there is a built-in filter capacitor that serves to smooth out the rectified pulses. This capacitor, of course, would eliminate our ability to detect when the AC cycle begins.

Circuit operation is straightforward. Since the LED only conducts electricity in one direction, it is "off" for half the time, and "on" for half the time of the cycle. In effect, the LED will be blinking 60 times per second (60 Hz).

Each time the LED blinks it causes the phototransistor to conduct (sink to ground). This part of the circuit operates like a standard optoisolator. The "sine-wave blinks" are squared up by the Schmitt-trigger and, subsequently, detected by the microcontroller.

This type of circuit could provide a time base reference, or trigger timing information for a "zero-cross" switching circuit.

Chapter 4

Chapter 5
Output

Chapter 5

Output

DC Motor

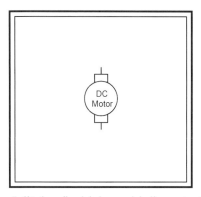

DC motors *can* be operated from the same power source as the other electronic circuitry, but in general it's not a very good idea.

This is because inductive devices (which a motor is), have the tendency to create "spikes" or "glitches" which could disrupt other, more sensitive circuits.

Figure 5.1 is a schematic that uses separate power supplies for the microcontroller and motor drive circuit. The interface driver between the two portions of the circuit "operates" on +5 volts (the supply voltage for the micon), but its output can be tied to a higher and separate voltage.

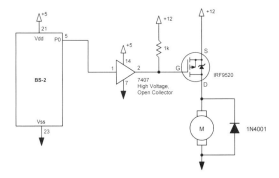

Figure 5.1
On/off DC motor control with a P-channel MOSFET

Chapter 5

Code 5.1
```
here:
    low 0              'turn the motor on for 1 second
    pause 1000
    high 0             'turn the motor off for 4 seconds
    pause 4000
    goto here
```

Both power supplies must have a common ground.

The circuit "sources" power to the motor through a P-channel MOSFET. Normally, it may be more convenient to drive a DC motor by tying one side to + voltage and then switching the motor on and off by connecting it to ground through an N-channel type device. N-channel devices have the advantage of being more readily available and are lower cost as well.

However, in certain situations – automotive applications for example – it may not be possible to "control" the grounding connection. In a typical car's electrical, most motors or electrical devices are "chassis" grounded – it's the + voltage that switches on and off.

Figure 5.2 adds a level of safety from voltage spikes by providing optical isolation. Voltage supply lines and grounds are completely separated.

Output

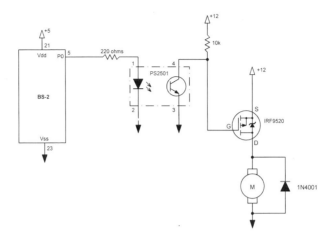

Figure 5.2
Simple optoisolated DC motor control

Code 5.2
```
here:
    high 0          'turn the motor on for 1 second
    pause 1000
    low 0           'turn the motor off for 4 seconds
    pause 4000
    goto here
```

Figure 5.2 also has the advantage of being "fail-safe." In the event of a signal line break in the control half of the circuit, phototransistor (contained inside the optoisolator) is in its "off" state. Therefore the 10K resistor is pulling the gate of the P-channel MOSFET high, which keeps the motor "off."

Figure 5.3 is designed to drive a very small, low current "hobby" type motor. Be sure and check the current requirements of the motor and the specifications of your transistor of choice for this circuit.

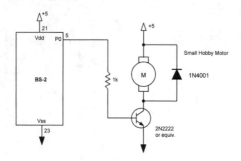

Figure 5.3
Small hobby DC motor control

Code 5.3
```
here:
high 0            'turn the motor on for 1 second
pause 1000
low 0             'turn the motor off for 4 seconds
pause 4000
goto here
```

Figure 5.4 is a PNP version of the prior circuit. Notice that we've added an inverter (74HC04) to drive the PNP type transistor. This provides for "positive-logic" control of the motor. In other words, a "high" output on P0 is inverted through the 74HC04, thus turning "on" the motor.

Again, be sure and check the specifications for both the motor and the specific transistor you choose for your application.

Output

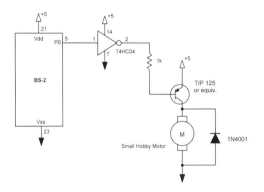

Figure 5.4
Small DC hobby motor control with PNP transistor

Code 5.4
```
here:
high 0              'turn the motor on for 1 second
pause 1000
low 0               'turn the motor off for 4 seconds
pause 4000
goto here
```

Figure 5.5 uses a MOSFET to drive an inexpensive relay. The motor *control* circuit is completely isolated from the motor *drive* circuit, as each section operates on completely separate power sources.

Relays like this are simple to use, but of course, are mechanical devices that eventually do wear out.

Chapter 5

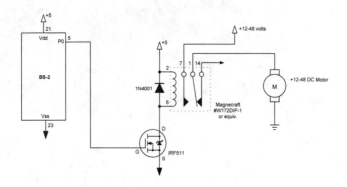

Figure 5.5
Hi-voltage DC motor control using a relay

Code 5.5
```
here:
high 0              'turn the motor on for 1 second
pause 1000
low 0               'turn the motor off for 4 seconds
pause 4000
goto here
```

The circuit in Figure 5.6 adds a high voltage inverter to drive the relay's 12-volt coil. You can add another inverter (in series) to provide "positive-logic" control, or you could use the 7407 non-inverting buffer-driver instead.

Output

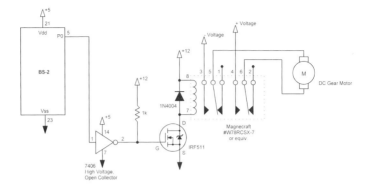

Figure 5.6
Relay control of a "+/-" power supply for a DC motor

Code 5.6

```
here:
    low 0           'turn the motor on for 1 second
    pause 1000
    high 0          'turn the motor off for 4 seconds
    pause 4000
    goto here
```

Figure 5.7 turns on a (small – less than 200 milliamps) DC motor for a period of time determined by the value of the 100K pot and the 10 uf capacitor.

A low going pulse on Pin 2 of the LM555 triggers the timing cycle for the motor. Pin 2 should be returned "high" immediately (after going "low") to ensure proper operation of the timing circuit.

Chapter 5

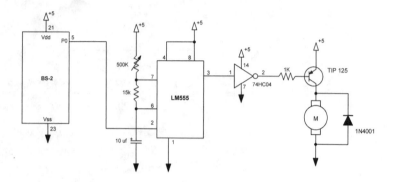

Figure 5.7
Pre-settable "time-on" DC motor control

Code 5.7
```
here:
low 0          'turn the motor on for a period of time
high 0         'set by the LM555 timer circuit

anywhere:

       'program can go on about its business
       'motor runs for duration set by
       'resistor and capacitors in LM555 circuit

goto anywhere
```

To increase the drive capacity of the above circuit, add an N-channel MOSFET, as shown in Figure 5.8. For higher current motors, be sure and use an appropriate heat sink for the transistor.

Output

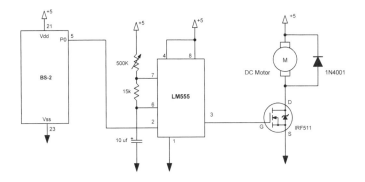

Figure 5.8
Hi-current pre-settable "time-on" motor control

Code 5.8
```
here:
   low 0              'turn the motor on for a period of time
   high 0             'set by the LM555 timer circuit

   anywhere:

      'program can go on about its business
      'motor runs for duration set by
      'resistor and capacitors in LM555 circuit

   goto anywhere
```

Figure 5.9 adds "timing cycle control" to the circuit. Now, under program control, the length of time that the DC motor is "on" can be set, via the solid-state pot (DS1804). Then, a negative going pulse (on pin 2 of the LM555) starts the cycle.

Chapter 5

This allows your code to set a pre-determined amount of time for the motor to run, start the motor, and then go on to "do other things." Your program doesn't need to worry about returning later to shut the motor off. It happens automatically.

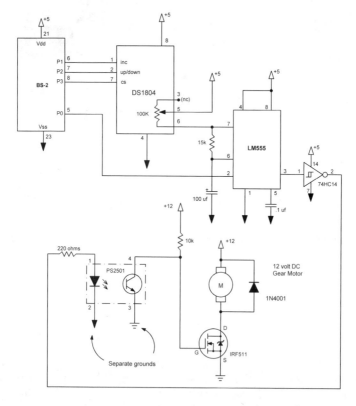

Figure 5.9
Micon adjustable "time-on", optically isolated DC motor circuit

Output

Code 5.9

```
x var byte
y var word

low 3              'select DS1804-010
low 2              'set direction to counter-clockwise
for x=1 to 100     'reset the pot to "zero"
high 1
low 1
next

high 2             'set direction to clockwise
for x=1 to 50      'set pot halfway - approx. 50k

high 1
low 1
next
high 3

here:
low 0              'turn the motor on for a period of time
high 0             'set by the LM555 timer circuit

anywhere:

    'program can go on about its business
    'motor runs for duration set by
    'resistor and capacitors in LM555 circuit

goto anywhere
```

If your project requires the motor direction to be reversible (as well as having a "pre-settable" time duration), then you could use the circuit shown in Figure 5.10.

The circuit uses a standard "H-bridge" MOSFET configuration, LM555 pre-settable timer circuit, and a digital potentiometer.

Chapter 5

"On-off" control is via pin 3 on the LM555 timer. Direction of rotation is controlled by the P4 and P5 I/O lines from the microcontroller.

Although the power supply circuits are separate (+5 and +12 volts respectively), their grounds must be connected together.

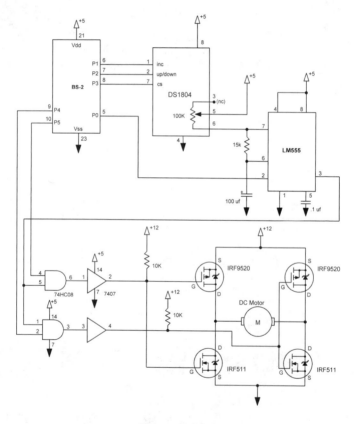

Figure 5.10
H-bridge adjustable "time-on" motor circuit

Code 5.10

```
    x var byte
    y var word

    low 3               'select DS1804-010
    low 2               'set direction to counter-clockwise
    for x=1 to 100      'reset the pot to "zero"
    high 1
    low 1
    next

    high 2              'set direction to clockwise
    for x=1 to 50       'set pot halfway - approx. 50k

    high 1
    low 1
    next
    high 3

    high 4              'set direction CW
    low 5

    low 0               'trigger the on cycle
    high 0
    pause 5000

    low 4               'set direction to CCW
    high 5

    low 0               'trigger the on cycle
    high 0
    pause 5000

    anywhere:
        'program can go on about its business
        'motor runs for duration set by
        'resistor and capacitors in LM555 circuit
    goto anywhere
```

Chapter 5

Output

Adjustable Flasher

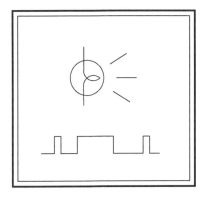

The LM555 can operate as a "stand-alone" blinker, as we've seen in other circuits. A single resistor connected to pin 7 and V+ can adjust its output timing.

By adding the digital potentiometer (DS1804) this timing cycle can be controlled by any micon.

Figure 5.11 incorporates an N-channel MOSFET for increased output drive current, as well as, "blink rate" control via the DS1804.

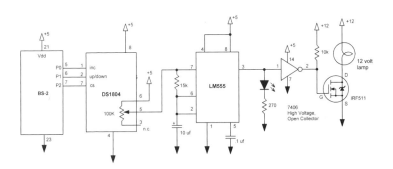

Figure 5.11
Hi-voltage, pre-settable lamp driver

167

Chapter 5

Code 5.11

```
    x var byte

    startover:
    low 2                  'select the '1804
    low 1                  'set direction to counter clockwise
    for x = 1 to 100       'tap it 100 times
    high 0
    low 0
    next                   'reset all the way counter clockwise

    high 1                 'set direction to clockwise

    here:
    for x= 1 to 100        'after each blink, increase the blink
    high 0                 'rate by one notch in the '1804
    low 0
    pause 150              'pause so we can see it
    next
    high 2
    high 0

    goto startover
```

You could add on/off control to the "automatic flasher" shown above by connecting pin 4 of the LM555 to another I/O line on the microcontroller. A "high" on this line enables the flasher, whereas, a "low" turns it off.

Figure 5.12 uses a low cost NPN transistor to drive a mechanical relay with a coil voltage of +12 volts. The relay will cycle automatically depending on the setting of the digital pot. Output current capability is dependent on the model of mechanical relay you choose.

Output

Because it's a relay, anything can be "automatically blinked" with this circuit – lights, motors, solenoids, and even air cylinders.

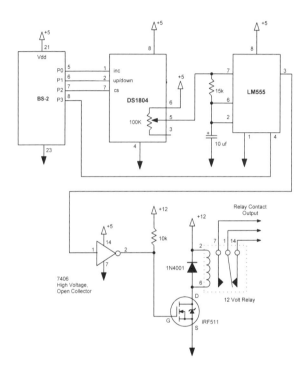

Figure 5.12
Adjustable "time-on" relay driver

Code 5.12
```
    x var byte

    startover:
    high 3          'enable LM555
```

Chapter 5

```
low 2                'select the '1804
low 1                'set direction to counter clockwise
for x = 1 to 100     'tap it 100 times
high 0
low 0
next                 'reset all the way counter clockwise

high 1               'set direction to clockwise

here:
for x= 1 to 50       'after each blink, decrease the blink
high 0               'rate by one notch in the '1804
low 0
pause 150            'pause so we can see it
next
high 2
high 0

low 3                'put LM555 into reset, shut off light driver
pause 4000
goto startover
```

Output

Audio Amplifier

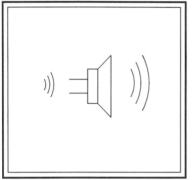

Audio amplifiers can have their gain adjusted automatically by including a digital pot in the circuit.

Figure 5.13 uses an LM386 audio amplifier coupled with a DS1804 digital potentiometer.

Audio output levels are controlled by the value of the pot which is set by the microcontroller. See Code 5.13.

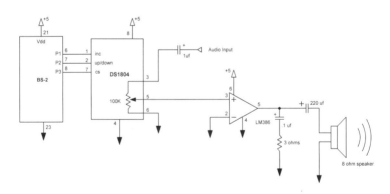

Figure 5.13
Volume control of an audio amplifier

Chapter 5

Code 5.13
```
    x var byte
    y var byte
    low 3              'enable the DS1804
    low 2              'set direction for CCW

    for x=1 to 100     'pulse 100 times
    high 1
    low 1
    next

    high 2             'set direction for CW

    y=14               'y = volume setpoint

    for x=1 to y       'between 1 and 100
    high 1
    low 1
    next

    there:             'once volume setpoint is reached
    goto there         'go do something else
```

An interesting application might be to hook up a microphone to this circuit. The device could monitor the sound level through your program and adjust it automatically, so as to prevent the sound from being "too loud." See *Cookbook Volume 1* for circuit samples.

Figure 5.14 uses the same LM386 audio amplifier. Operation (volume control) is similar to the above schematic. However, this circuit adds another feature to your "audio control" project – "mute."

Output

Since the "on-state resistance" of the N-channel MOSFET is very low, it can be used as a "switch" in the ground leg connection of the speaker itself.

A "high" on P3 enables sound through the amplifier, and conversely, a "low" mutes it – regardless of the volume setting within the DS1804 potentiometer.

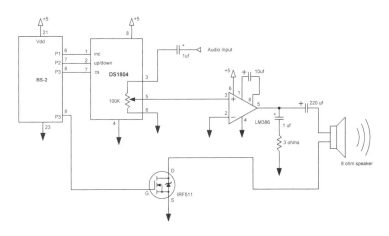

Figure 5.14
Volume control with "muting" capability

Code 5.14

```
x var byte
y var byte
low 3                   'enable the DS1804
low 2                   'set direction for CCW

low 4
for x=1 to 100          'pulse 100 times
high 1
low 1
next
```

Chapter 5

```
high 2              'set direction for CW

y=20                'y = volume setpoint

for x=1 to y        'between 1 and 100
high 1
low 1
next

there:
high 4
pause 1000
low 4
pause 1000          'once volume setpoint is reached
goto there          'go do something else
```

Watchdog Timer

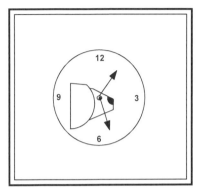

A watchdog timer is a circuit that acts as an "overseer" or "fail-safe" device.

Let's suppose you have a project that has to accomplish a certain sequence of tasks, loop back to the beginning, and start all over again. If the hardware and code always operate correctly, then everything is fine.

However if something causes your program to "hang", what then? Does your creation just sit there waiting until you notice its non-functionality? Probably so.

A watchdog timer expects a periodic *"I'm OK and still functioning properly"* signal from the brain – in our case, usually a microcontroller.

If the watchdog does not receive this signal (typically a pulse) within a certain finite time period, it creates an output indication that can reset or re-start the circuitry.

Figure 5.15 configures an LM555 timer as a watchdog device. As long as pin 7 (on the 555) receives a pulse within a given time period (determined by the 1 meg resistor and the capacitor connected to pin 2), its output remains high.

Chapter 5

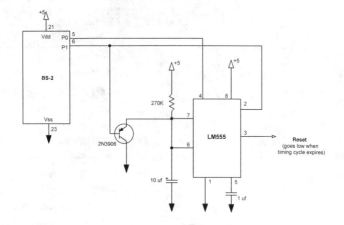

Figure 5.15
LM555 watchdog timer circuit

Code 5.15

```
x var byte
high 0

here:
for x = 1 to 100      'send a string of regular pulses
high 1                'spaced at a maximum amount
pause 5               'of time set by the resistor and
low 1                 'capacitor in the 555 timer circuit
pause 5
next
high 1                'take the trigger line back high
pause 5000            'and leave it there
goto here             'output of 555 goes low with no
                      'incoming pulses
```

But if the expected pulse does not arrive in time, then the output of the LM555 goes low. This low output can be connected to an LED or other device, indicating a problem.

Another feature of Figure 5.15 is that the micon can instigate a "reset" signal by causing P0 to pulse "low." This resets the LM555 timer, causing the output on pin 3 to go low, initiating a re-start sequence within your circuitry.

Refer to Figure 5.16. By connecting a resistor and capacitor to the output of the LM555 and attaching that point to the reset line (which is "active low") on the Stamp, your program must successfully complete its mission. If it doesn't, then the Stamp will perform a "reset" and "reboot" your program so that the problem can "clear itself", giving it another try.

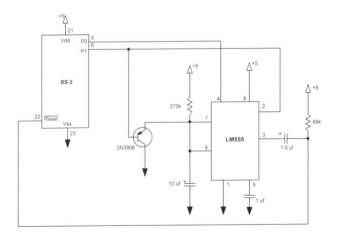

Figure 5.16
Watchdog reset circuit for the BASIC Stamp 2

Chapter 5

Code 5.16

```
    x var byte
    high 0

    debug "If you can read this, the micon has been reset"
    debug cr
    pause 3000

    here:
    debug "The LM555 is receiving pulses now..."
    debug cr

    for x = 1 to 100      'send a string of regular pulses
    high 1                'spaced at a maximum amount
    pause 5               'of time set by the resistor and
    low 1                 'capacitor in the 555 timer circuit
    pause 5
    next
    high 1                'take the trigger line back high
    debug "The pulses have stopped!!!"
    debug cr

    pause 5000            'and leave it there
    goto here             'output of 555 goes low with no
                          'incoming pulses
```

In Figure 5.17 we've implemented a "re-triggerable one-shot" utilizing the 74LS123. As long as the 74LS123 continues to receive a series of pulses (within a time frame determined by the resistor and capacitor), the LED remains off.

If your program fails to return in time to deliver the "I'm OK" pulse, the LED comes on. Of course, you can connect this output to additional circuitry and initiate whatever "reset" tasks your project requires.

Output

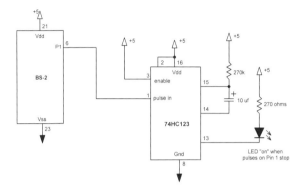

Figure 5.17
74HC123 watchdog timer with LED indicator

Code 5.17

```
x var byte
here:
low 0
for x = 1 to 100
high 1           'a solid stream of pulses to
pause 5          'retrigger the '123
low 1
pause 5
next

debug "no pulses now..."
debug "the LED will come on now because"
debug "the pulses have stopped for more than 4 seconds"
pause 4000
high 0
goto here
```

Chapter 5

Figure 5.18 adds a relay driver to the output on pin 4 of the 74HC123. This provides greater drive capability, as well as, circuit isolation between the control and reset portions of the circuit.

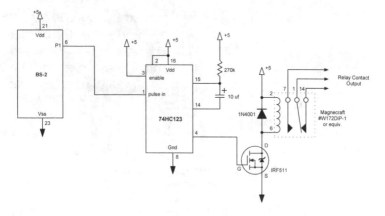

Figure 5.18
Relay output watchdog circuit

Code 5.18

```
      x var byte
      here:
      debug "a solid stream of pulses..."
      debug cr
      for x = 1 to 100
      high 1            'a solid stream of pulses to
      pause 20          'retrigger the '123
      low 1
      pause 20
      next

      debug "no pulses now..."
      debug cr
      debug "the Relay will turn on now because"
      debug cr
```

Output

```
debug "the pulses have stopped for over 4 seconds"
debug cr
pause 4000
goto here
```

A "latching reset" function is performed in Figure 5.19. If the 74LS123 fails to receive the required pulse, the output on pin 4 turns "on" the first transistor. That transistor energizes the double-pole, double-throw relay which has one set of contacts connected in such a way as to turn "on" the other transistor.

With the second transistor in the "on" state, the relay is actuated no matter what the status of the 74LS123.

In order to bring the relay contacts out of a "reset situation", the normally closed push button must be activated.

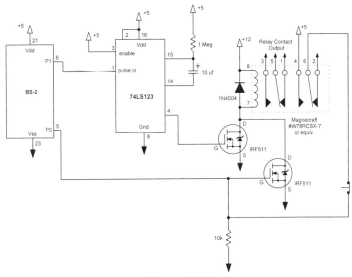

Figure 5.19
Latching output watchdog relay circuit

Chapter 5

Code 5.19

```
x var byte

startover:
debug "a solid stream of pulses..."
debug cr
for x = 1 to 100
high 1              'a solid stream of pulses to
pause 20            'retrigger the '123
low 1
pause 20
next

debug "no pulses now..."
debug cr
debug "the relay will turn on and stay on"
debug cr
debug "until someone pushes the switch"
debug cr
pause 4000

there:
if in0=0 then startover
goto there
```

Stepper Motor

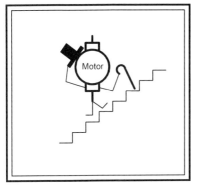

Stepper motors are found almost everywhere, from personal computer disc drives to portable compact disc players. They're also quite inexpensive – as low as a couple of dollars in some cases.

A stepper motor requires a precisely controlled series of pulses, applied in the right sequence, amongst several different coil inputs, to operate properly.

Yes, a microcontroller can produce the signals required to drive a stepper. In fact, there are products that you can purchase right now from companies that have taken the time to create and program such devices.

Figure 5.20 utilizes the EDE1200. It's been specifically designed for the control of stepper motors. The chip can be easily interfaced to your micon of choice.

This circuit utilizes a "unipolar" type stepper motor and is driven by some high current MOSFET's. The circuit is set up for "Run" mode, which means that the stepper motor will spin automatically, at a speed determined by the three bit binary value presented on pins 11, 12 and 13.

Chapter 5

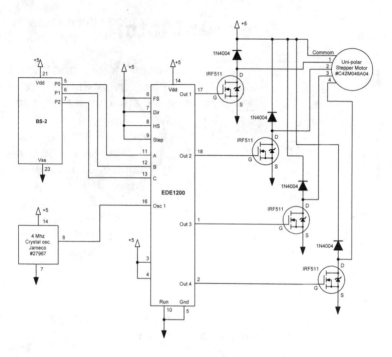

Figure 5.20
Stepper motor controller circuit

Code 5.20

```
x var byte
dira=7              'set up (3) speed bits as outputs
here:
for x = 0 to 7      'go from slow to fast
outa=x
pause 1000          'over a 7 second time period
next
for x = 7 to 0      'fast to slow
outa=   x
pause 1000
next
goto here
```

Output

Figure 5.21 utilizes a ULN2003A integrated circuit that has four independent drivers contained within the same package.

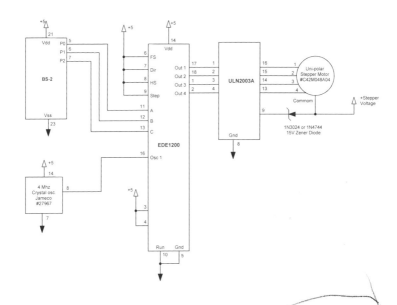

Figure 5.21
Stepper motor control using the ULN2003A

Code 5.21
```
low 0           'set speed and forget about it
high 1          '000=slow, 111= fast
high 2          'this speed setting = 011

here:
                'the rest of your program here
goto here
```

Chapter 5

D/A Conversion

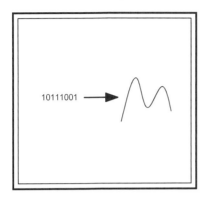

Going from a digital value to an analog representation of that value can be accomplished in a multitude of ways.

MAC Volume 1 has a simple "resistive ladder" D/A converter circuit, but it may not be suitable for more precise applications.

Figure 5.22 utilizes an AD558. This chip is specifically designed to deliver a *very* accurate analog representation of a digital value. Internally, it's been "laser trimmed" for accuracy at the factory.

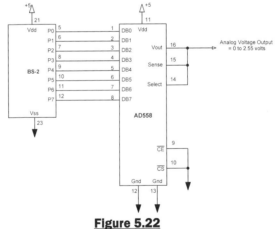

Figure 5.22
High accuracy D/A converter
0 – 2.55 volts output

Chapter 5

Code 5.22
```
    x var byte

    here:
    dirl=255            'set P0-P7 to outputs
    for x= 0 to 255
    outl=x              'step output 255 times
    pause 100           'every 1/10th of a second
    next                'with a full scale output of 2.55 volts
    pause 3000

    goto here
```

The output voltage on the AD558 ranges from 0 to 2.55 volts over an 8-bit digital value input. That is equal to .01 volts per step (8 bits = 255, plus the value of "zero" for a total of 256 "steps" of resolution).

Figure 5.22 is not the most efficient use of I/O lines, but the schematic and Code are quite easily implemented.

Figure 5.23 drops the number of required I/O lines down to four, and yet still maintains 8-bit accuracy. This is accomplished by using a serial to parallel converter (74HC164).

Output

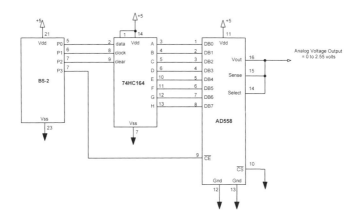

Figure 5.23
A/D converter with "set and forget" capability
0 – 2.55 volts output

Code 5.23

```
x var byte

high 3              'disables the AD558

here:
low 2               'clears the '164
high 2

x=213               'send "213" to the '164

shiftout 0,1,msbfirst,[x]

low 3               'once all 8 bits have "ripple"
high 3              'into the '164, strobe them in
pause 2000          'parallel into the AD558

there:              'output of AD558 = 2.13 volts
goto there
```

Chapter 5

As shown in Figure 5.24 increasing the supply voltage on the AD558 to +12 volts, and connecting "select" to Vout, we now have an analog voltage output span capability 0 to +10 volts.

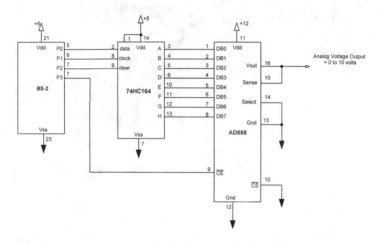

Figure 5.24
*Serial to parallel D/A converter with
0 – 10 volt full-scale output*

Code 5.24
```
    x var byte

    high 3              'disables the AD558

    here:
    low 2               'clears the '164
    high 2

    x=128               'send "128" to the '164

    shiftout 0,1,msbfirst,[x]
```

Output

```
low 3              'once all 8 bits have "ripple"
high 3             'into the '164, strobe them in
pause 2000         'parallel into the AD558

there:             'output of AD558 = 5.00 volts

goto there
```

This results in about .0395 volts for each digital step. (.0395 volts times 255 steps = 10.07 volts). The circuit requires two separate power supply voltages.

Chapter 5

Output

Voice Record and Playback

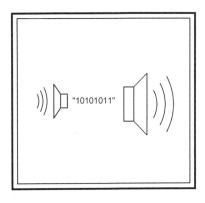

Digital answering machines and other sound recorder systems take advantage of certain "single chip" solutions that we can interface to a microcontroller.

Figure 5.25 utilizes the ISD1110 integrated circuit. Designed to record and playback audio with all "solid-state" technology. Solid-state technology means there is no tape; sound is stored digitally within the chip itself.

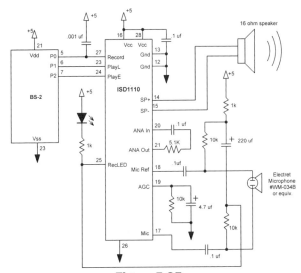

Figure 5.25
Micon controlled sound record and playback circuit

Chapter 5

Code 5.25
```
    here:

    high 1              'disable level triggered playback
    high 2              'disable edge triggered playback

    pause 2000          'wait for 2 seconds before recording
    low 0               'begin record cycle for 3.5 seconds
    pause 3500
    high 0

    pause 1000          'wait for 1 second before playing back
    low 2               'trigger edge playback mode
    high 2

    pause 3500          'playback for 3.5 seconds

    goto here
```

The ISD1110 has a total of ten seconds of record/playback time.

Rather than connect this device to some push-button switches for manual operation, connecting to a micon provides for a completely automated sound record and playback system.

Bar Graph Driver

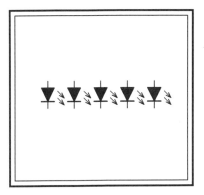

Bar graph drivers are cool. I admit it. As simplistic as it sounds, watching those LED's go up and down, or back and forth, is almost addictive.

Bar graph drivers are really designed as "stand-alone" devices. However, when coupled with a microcontroller, they can take on a whole new dimension.

The LM3914N is designed to interface directly to ten separate LED's. You could also use one of those LED arrays that consists of ten rectangular LED's all contained within the same package. They're quite convenient.

Figure 5.26 is perhaps the simplest implementation of a bar graph driver. Using an analog voltage, generated by the Stamp and R/C converter circuit, the LM3914N is in "bar display" mode. If you prefer "dot" mode, leave pin 9 floating, that is, not connected to anything.

The voltage input range for this circuit is 0-5 volts DC, but in order for the chip to operate properly, the supply voltage of the LM3914N must be at least 6.8 volts. In this circuit we're using a 9-volt source. Check out Chapter 2 of this Cookbook for different ways to achieve multiple voltage output power supplies.

Chapter 5

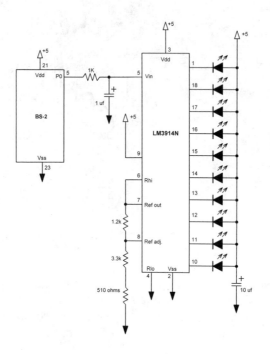

Figure 5.26
PWM D/A converter driving an LED bargraph

Code 5.26

```
x var byte

here:

for x = 1 to 255

pwm 0,x,30        'output analog voltage
pause 10          'slowly ramping up
next
```

```
    pause 20000      'monitor the voltage as it
                     'bleeds from the capacitor

    low 0            'drop the cap voltage to zero
                     'and shut all lights off

    pause 2000

    goto here
```

Figure 5.27 incorporates the use of a solid-state potentiometer to deliver the proper amount of analog voltage for subsequent display on the bar graph.

This allows the microcontroller to "set and forget" the voltage level, allowing it to go attend to other matters related to its code execution. Meanwhile the bar graph display remains constant, until modified by the program.

Chapter 5

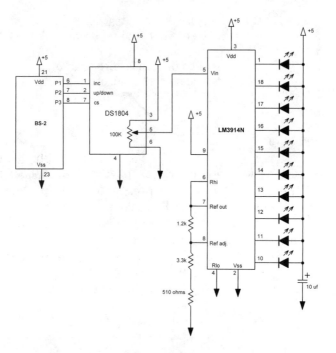

Figure 5.27
"Set and forget" LED bargraph driver

Code 5.27

```
x var byte

low 3             'select DS1804
low 2             'set direction to counter-clockwise
for x=1 to 100    'reset the pot to "zero"
high 1
low 1
next

high 2            'set direction to clockwise
for x=1 to 50     'set voltage for 2.5 volts
```

Output

```
    high 1         'toggle the pot
    low 1
    next

    here:
                   'rest of program goes here
       goto here
```

"Cascading" or connecting multiple bar graph drivers together is shown in Figure 5.28. Full-scale readout is from about zero to 2.5 volts DC. The circuit is in "dot" mode.

The voltage input is derived from the (two) 1K-resistor divider network connected to +5 volts. The resultant 2.5 volts is then input into the digital pot for subsequent control via the microcontroller.

Chapter 5

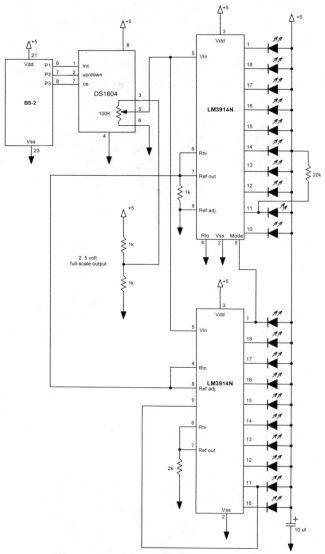

Figure 5.28
Cascaded LED bargraph driver circuit

Output

Code 5.28
   x var byte

```
low 3                   'select DS1804
low 2                   'set direction to counter-clockwise
for x=1 to 100          'reset the pot to "zero"
high 1
low 1
pause 50                'over a 5 second time period
next

high 2                  'set direction to clockwise
for x=1 to 100          'set voltage for 2.5 volts
high 1                  'toggle the pot
low 1
pause 100               'over a 10 second time period
next
```

Figure 5.29 incorporates a serial to parallel converter, as well as a "resistive ladder" D/A converter circuit. By tying pin 9 (on the LM3914N) to +voltage, the display driver is in "bar" mode.

With each of these circuits, it is important to keep all of the ground connections as short as possible. As you assemble the circuit, try to have the ground connections all tie into a single junction as near to pin 2 (on the LM3914N) as is possible. This will help minimize oscillations and glitches throughout the system.

Chapter 5

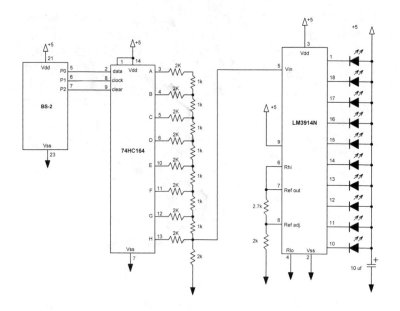

Figure 5.29
Serial to parallel D/A converter driving
the LM3914N

Code 5.29
```
x var byte

here:
low 2              'clears the '164
high 2

for x=0 to 255     'pick any number 0-255

shiftout 0,1,msbfirst,[x]
pause 40
next               'watch the LED's for "ripple" effect

goto here
```

Output

Figure 5.30 incorporates a latch into the above circuit, which prevents the "ripple effect" when loading data into the 74HC164. By using the enable line (pin 11 on the 74HC374), the serial to parallel converter can be loaded with a new value. Then all eight bits are "latched" simultaneously through the '374 to the D/A conversion circuit.

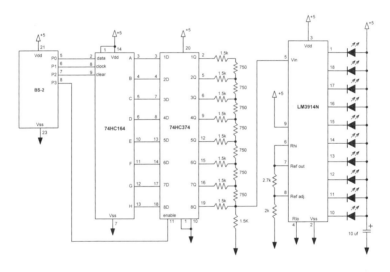

Figure 5.30
Latched D/A converter driving the LM3914N

Code 5.30
```
    x var byte

    low 2                      'clears the '164
    high 2
```

Chapter 5

```
for x=0 to 255            'full scale reading (3.0 volts)
low 3
shiftout 0,1,msbfirst,[x]  'send out 8 bits
high 3                    'latch the data into the '374
next
pause 1000

for x=0 to 128            'half scale reading
low 3
shiftout 0,1,msbfirst,[x]
high 3                    'latch the data into the '374
next

here:

goto here
```

An expanded version of the above circuit is shown in Figure 5.31. The circuit is in dot mode and will display up to 2.5 volts full-scale readout.

Output

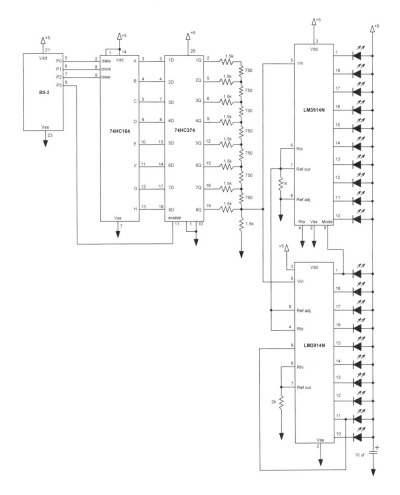

Figure 5.31
Cascaded LM3914N bargraph circuit with input latch

Chapter 5

Code 5.31

```
x var byte

low 2                     'clears the '164
high 2

here:
for x=0 to 255            'full scale reading (3.0 volts)
low 3
shiftout 0,1,msbfirst,[x] 'send out 8 bits
high 3                    'latch the data into the '374
next
pause 100

for x=255 to 0            'half scale reading
low 3
shiftout 0,1,msbfirst,[x]
high 3                    'latch the data into the '374
next

goto here                 'bounce back and forth
```

Chapter 6
System Interfacing

Chapter 6

System Interfacing

RS-232

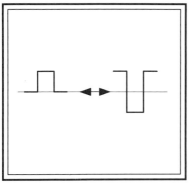

Data transmission from one circuit to another can be accomplished via a serial connection. "Serial" implies that the data is "all in a line", which is how the bits are sent – one at a time, all in a row.

The RS-232 "standard" attempts to standardize on a hardware specification allowing separate electronic systems to communicate over a serial connection.

Simply put, RS-232 uses positive and negative voltage swings to indicate a "1" or "0". The positive and negative voltages are present on a single signal line, and are measured and referenced to a common ground connection between the two inter-connected systems (transmitter and receiver).

Figure 6.1 uses two of the most widely available chips for implementing an RS-232 connection.

The primary drawback of the 1488 and 1489 "line driver" and "line receiver" chips is the requirement for both positive and negative power supplies (for the MC1488 specifically).

Refer to Chapter 2 of this *Cookbook* for some easy to implement negative power supplies.

Chapter 6

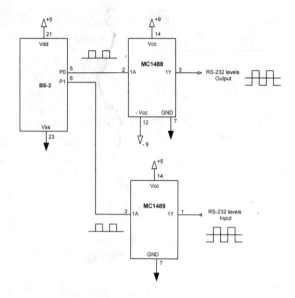

Figure 6.1
0 – 5 signal conversion to RS-232 voltage levels

Code 6.1
```
here:
high 0      'data output
low 0
goto here
```

Figure 6.2 uses the MAX232CPE RS-232 line driver/receiver from Maxim. The device generates its own negative voltage using just a couple of external capacitors, thereby eliminating the need for a separate external "minus" supply.

System Interfacing

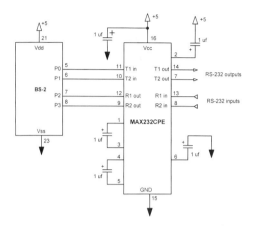

Figure 6.2
RS-232 level conversion with single voltage supply

Code 6.2
```
here:
    high 0
    low 0        'pulses of data
    goto here
```

Figure 6.3 uses the MAX233CPP, which can generate the required negative voltage swing without capacitors.

Chapter 6

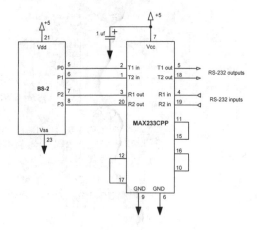

Figure 6.3
RS-232 conversion with single voltage supply and no capacitors

Code 6.3
```
here:
    high 0      'could be serial output command
    low 0
    goto here
```

Entire books can be written on the subject of serial communications, and in fact several have been. Check out the "Resources" section in this *Cookbook*.

System Interfacing

RS-485/RS-422

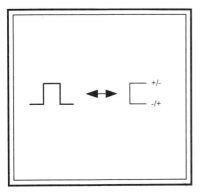

Unlike RS-232, which relies on a voltage swing from positive to negative (in relation to ground) to indicate a "1" or "0", the RS-485 / RS-422 standard relies on a "balanced" pair of data transmission lines.

The voltage potential produced between these two balanced lines alternates polarity in relation to each other, *not* in relation to ground.

The primary difference between RS-422 and RS-485 is that RS-485 implementation *requires* an "enable" control signal - whereas RS-422 *may* have an enable signal, but doesn't necessarily have to. Both systems use the "balanced line driver" methodology.

Figure 6.4 uses the MAX485 integrated transceiver. Data is transmitted from the Stamp into pin 4 on the MAX485 when DE (Driver Enable) is high. Data is received out of pin 1 (on the MAX485 when RE (Receiver output Enable) is low. Therefore in this circuit, P1 on the Stamp is acting as a data direction control line.

Chapter 6

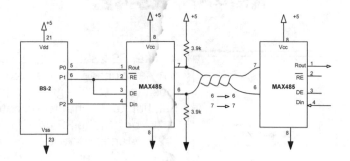

Figure 6.4
RS485 / RS422 level conversions

Code 6.4
```
here:
high 1      'enable output driver
high 2      'data transfer out
low 2
low 1       'receive enabled
            'serial input routine goes here
goto here
```

Figure 6.5 illustrates how several (actually up to 128) MAX485 transceivers can be connected along the same twisted pair cable.

Transmission length is up to 4000 feet.

System Interfacing

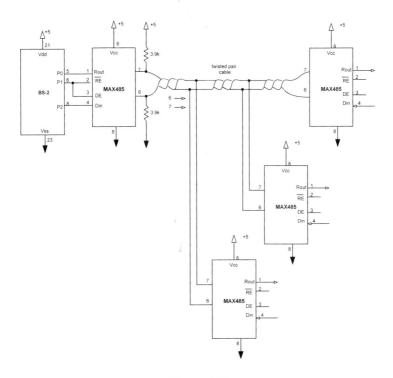

Figure 6.5
"Daisy chained" RS-485 circuit

Code 6.5

```
here:
    high 1      'enable output driver
    high 2      'data transfer out
    low 2
    low 1       'receive enabled
                'serial input routine goes here
    goto here
```

An alternative to the Maxim MAX485 is shown in Figure 6.6. The DS96174N transmitter contains four driver circuits

Chapter 6

within the same package. The DS96175N receiver has four receiver circuits as well.

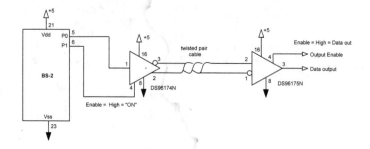

Figure 6.6
RS-485 conversion using the DS96174N and DS96175N

Code 6.6
```
here:
high 1                  'enabled output

high 0                  'data out
low 0

low 1                   'disabled output
goto here
```

Figure 6.7 is a bi-directional version of the above schematic. Independent transmitter and receiver chips are required at each end.

System Interfacing

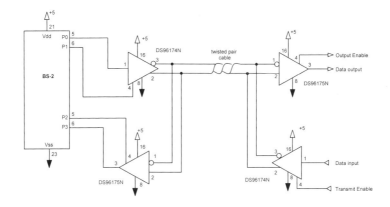

Figure 6.7
Bi-directional circuit using the DS96174N and DS96175N

Code 6.7
```
    x var bit

    here:
    high 1              'enabled output

    high 0              'data out
    low 0
    low 1               'disabled output

    high 2              'enable receiver
    x=in3               'get input bit
    goto here
```

Transmission speeds up to 10 Mb/s are possible, as well as, distance capability of up to 4000 feet from end to end. With proper handshaking and careful coding, you could use the same pair of I/O lines for both transmit and receive functions.

Chapter 6

System Interfacing

Remote control

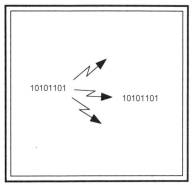

Where would we be without the TV remote? Remote control has done more for the "lazy" in us than all comfy recliners combined.

This first circuit is an "encoder". This is a circuit that creates a sequence of data that can then be connected to a "transmission medium" such as infrared or radio wave.

Figure 6.8 uses the HT640 encoder chip. This device is specifically designed for use in such applications as garage door openers, burglar alarm systems, and cordless telephones.

The chip consists of 18 address and/or data inputs. Figure 6.8 depicts the chip in one of its more common configurations. A 10-position dipswitch is used to determine the address of the chip itself. This means that unless the address of the transmitter matches the address on the receiver (also preset by a dipswitch), the data bits to follow will be ignored.

Chapter 6

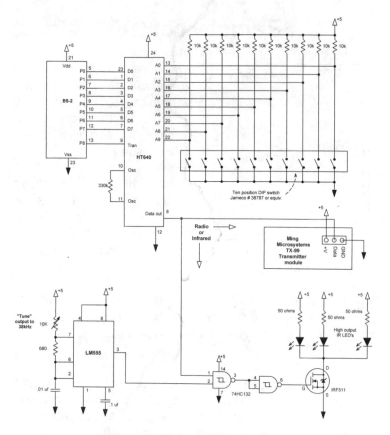

Figure 6.8
Encoder / transmitter circuit using a 10-bit address

Code 6.8

```
x var byte
dirl=255            'set up P0-P7 as outputs
low 8               'disable transmit function

here:
for x=1 to 25
```

System Interfacing

```
    outl=255    'send 11111111" 25 times "through the air"
    high 8
    low 8
    pause 50
    next

    for x=1 to 25
    outl=0      'send "00000000" 25 times "through the air"
    high 8
    low 8
    pause 50
    next

    for x=1 to 50
    outl=170    'send "10101010" 50 times "through the air"
    high 8
    low 8
    pause 50
    next

    goto here
```

Data output can go to either the infrared transmitter or to the radio transmission circuitry.

The infrared circuit is "homebrewed" with common components. The radio circuit is "store bought". It's inexpensive and works quite well for its price.

Chapter 6

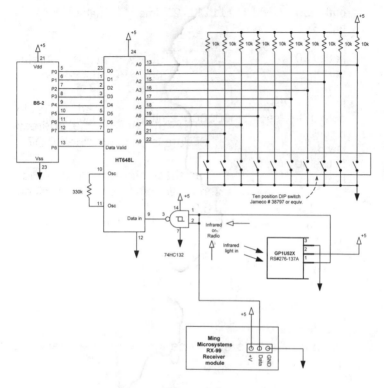

Figure 6.9
Receiver / decoder circuit using a 10-bit address

Code 6.9

```
    x var byte

    here:
    if in8=1 then get_data
    goto here

    get_data:
    x=inl                   'get the incoming data bits
    debug ? x               'and show me
```

      debug cr
      goto here

Figure 6.9 is the complementary "decoder circuit." Again, the 10-position dipswitch sets the address of this device.

"Demodulated" data (from either the infrared or radio receiver) is input into the HT648L, and if the address information matches, the 8-bit data is presented on D0-D7.

When valid data is available on these lines, the "data valid" line goes "high". This is used as an indication for the micon to read the byte. See Code 6.9.

Chapter 6

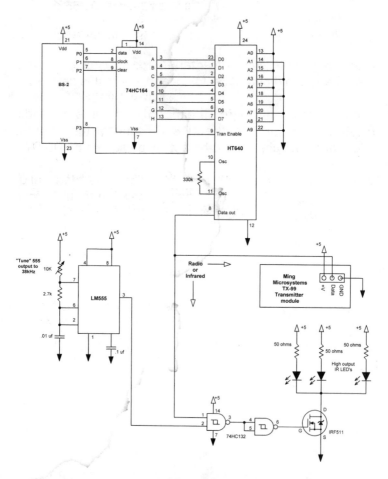

Figure 6.10
Encoder circuit that requires fewer I/O lines

Code 6.10

```
x var byte
here:
low 2                    'clears the '164
high 2
```

System Interfacing

```
x = 159                      'pick any number 0-255

shiftout 0,1,msbfirst,[x]    'load "159" into the '640

high 3                       'transmit it
low 3
pause 50                     'give it time to ripple out

goto here
```

Figure 6.10 depicts another way to set the address of both the encoder and transmitter. Instead of using a dipswitch, tie high or low any combination of A0-A9. Just be sure that both encoder and decoder match.

As you can see, no pull-up resistors are required, the address lines are simply bonded to +V or ground.

In fact, you could leave any address pin "open" (for a "third" state) and greatly increase the number of addresses possible.

Chapter 6

Resources

   1-888-512-1024
www.parallaxinc.com
www.stampsinclass.com

The source for Stamp related stuff. Many of the parts used in this book are available here as well.

JAMECO   1-800-831-4242
ELECTRONICS
www.jameco.com

Extensive selection of Electronic components as well as computer parts. Free catalog. Jameco Part numbers for many of the components used in this Volume are listed at the end of this section.

Mouser Electronics   1-800-346-6873
www.mouser.com

Comprehensive selection of electronic components. Free catalog on request. Entire catalog available on website. No minimum order requirements and they ship same day with no handling charges. Complete tech and quote department – provides cross-referencing services and specification sheets.

Resources

Allied Electronics   1-800-433-5700
www.alliedelec.com

Another source for electronic components. Free catalog and CD-ROM on request.

Digi-Key   1-800-344-4539
www.digikey.com

Another basic electronic component parts source. Free catalog.

Radio Shack   1-800-THE-SHACK
www.radioshack.com

They're everywhere. Basic selection of component parts at their retail stores. Through "Radio Shack Unlimited", they have a much wider selection, and will ship to your door.

Edmund Scientific   1-856-573-6250
www.edmundscientific.com

Nifty devices for optical applications (lenses, cameras, lasers, etc.)

WM Berg   1-800-232-BERG
www.wmberg.com

All kinds of mechanical drive components, etc.

Resources

McMaster-Carr   1-630-833-0300
www.mcmaster.com

An unbelievable assortment of mechanical devices – motors, solenoids, etc. Catalog has an extensive index. Call the above number for their local distribution center and phone number.

Other information sources:

Check out my website at **www.miconcookbook.com** for information and updates to this book.

Of course, no micon library is complete without the original ***Microcontroller Application Cookbook*** ☺ "The beginners sourcebook for real world machine control."

ISBN: 0-615-11552-7
Available at bookstores and websites everywhere including www.parallaxinc.com.

For those of you who seek to protect your "*Cookbook* developments", I recommend ***From Patent to Profit.*** It's written by Bob DeMatteis (see Foreword in this *Cookbook*), and will significantly reduce headaches when deciding, "where do I go from here."

Resources

Whether you are a professional inventor, a part-time dabbler, or just a clever daydreamer, *From Patent to Profit* will help you turn your creative ideas into real money. This 5-star rated book shows how to develop your ideas and inventions into real products at little or no cost! Recommended by the U.S. Patent Office and endorsed by the Small Business Development Centers.

ISBN: 0-399-52738-9 Perigee/Penguin Putnam Press

It's available from http://frompatenttoprofit.com, numerous other websites including Amazon, as well as bookstores everywhere.

Stampworks, by Jon Williams
Parallax, Inc.; ISBN: 1928982077;
(October 1, 2000)
Tons of applications examples. The book is available separately or with a "kit" that contains all components and tools necessary for building and working with the multitude of projects outlined in the book.

BASIC Stamp Manual
Parallax, Inc.; ISBN: 192898200X;
A "must have" for any serious Stamp users. Loaded with descriptions of all Stamp commands as well as applications examples. Available at parallaxinc.com for free download, or for purchase in printed form.

Resources

Programming and Customizing the Basic Stamp Microcontroller, by Scott Edwards. Tab Books; ISBN: 0071371923; 2nd edition (March 21, 2001)

Microcontroller Projects Using the Basic Stamp, by Al Williams. CMP Books; ISBN: 1578201012; 2nd edition (February 2002)

Basic Stamp, by Claus Kuhnel, Klaus Zahnert Newnes; ISBN: 0750672455; 2nd edition (July 2000)

The National Electrical Code
ISBN 0-442-02223-9

Wiring Simplified, by Richter and Schwan
ISBN 0-9603294-4-7

Serial Port Complete, by Jan Axelson
(ISBN # 0965081923)

Serial PIC'n, by Roger Stevens.
(ISBN # 0965416224)

Parallax's website at www.stampsinclass.com has a large amount of free (downloadable) learning materials such as:

What's a Microcontroller?
Robotics
Environmental Projects
Analog and Digital

These materials provide a great foundation for exploring the world of the microcontrollers, especially if you're just starting out.

Resources

This is also where you can download a complete copy of the BASIC Stamp Manual.

Be sure to join the discussion forums here too. And if you're new to microcontrollers (especially the Stamp), don't be afraid to post questions. There are many kind people here who love to enlighten those of us who are interested in learning more about this technology.

Many of the components used in these Cookbook circuits are available from Jameco Electronics. I've included the manufacturer part numbers as well as their associated Jameco part numbers below for easy reference.

Mfr. Part #	Jameco Part #	Description
W01M	178132	Bridge
M/N40	184751	AC/AC trans. 450 ma.
AC1216M3	153761	AC/AC trans. 1600 ma.
AC1810F1	121216	AC/AC trans. 1000 ma.
DC120F1	100870	AC/DC trans. 1000 ma.
XFR122 / F44X	29225	Power Trans. 12v @ 2 amps
XFR124 / F181U	102120	Power Trans. 12 @ 4 amps
XFR242	99653	Power Trans. 24 @ 2 amps
MDA990-3	25574	Bridge
LM323K	23667	Regulator
PS2501-4	160338	Optoisolator
206-8LP	139011	8 pos dipswitch
74LS151	46703	8 input mux
74150	49509	16-1 mux
MX061	136557	Push switch
74HC165	45495	74hc165
ERD110RS	139635	10 pos rotary switch

Resources

Mfr. Part #	Jameco Part #	Description
IRD500	112168	Infrared det
TLN110	106526	Infrared diode
H21A1	114091	Opto interruper
FPM07PG	163731	Pressure sensor
3540S-1-103	183521	10 turn pot
74HC193	45559	74hc193
74HC688	46113	74hc688
EDE1144	171969	Keyboard enc
OSC4	27967	Crystal osc
H11AA1	18825	AC input isolator
S1805N1	155547	Ind. proximity sensor
LM331N	23721	Voltage tofreq.
LM2907N	23317	Freq. To voltage
LM380N	24037	2 watt audio amp
LM383T	24088	7 watt audio amp
EDE1200	141532	Stepper controller
AD558JN	115158	8 bit dac
ISD1110P	141671	Speech chip
ISD25120P	141655	Speech chip
WM-034B	189958	Microphone
XR-2206	34972	Function generator
LM3914N	24230	Bar graph display
MAX232CPE	24811	RS232 driver
MAX233CPP	106163	RS232 driver
HT640	126594	Encoder
HT648L	126607	Decoder
206-10	38797	10 pos dip switch

Resources

Index
(Volumes 1 & 2)

Numbers beginning with an "I" or "II" are from *The Microcontroller Application Cookbook, Vol. 1 and 2 respectively.*

120 VAC	I-109, I-111, I-178, II-19
1N4001	I-155
2N2222	I-42, I-155, II-69
3M	I-29
555	I-60, I-61, I-80, I-136, I-154, I-170, I-184, II-89, II-91, II-94, II-175, II-176, II-178
558-IAC5	I-111
7404	I-115
7406	I-150, I-178, II-69
7407	I-150, I-156, II-158
74HC04	I-115, I-116, II-66, II-156
74HC05	I-117
74HC08	I-123, I-125
74HC123	II-179, II-180
74HC138	I-204
74HC139	I-202, I-212
74HC14	I-72, I-107, I-160, II-90
74HC151	I-195, I-196, I-197, I-199, I-202, II-74
74HC154	I-203, I-204
74HC164	I-208, II-64, II-188, II-203
74HC241	I-119
74HC245	I-120
74HC374	I-210, I-211, II-64, II-67, II-203
74LS74	I-62

A
A/D	II-115, II-117, II-123, II-189
A/D converter	I-69, I-77, I-81, I-84, I-86, I-90, I-91, I-100, I-102, I-220, I-223, I-232, II-115, II-117, II-123, II-18

235

Index

AC	I-159, I-173, I-174, I-175, I-179, II-6, II-19, II-20, II-21, II-26, II-27, II-147, II-148, II-234, II-235
active high	I-57, I-58
active low	I-55, I-56, I-59, I-222, II-177
actuated	II-73, II-86, II-131, II-181
AD0831	I-77, I-90, I-102, I-221, I-223, I-224, I-225, I-227, I-228, I-232, I-233, II-123
AD558	II-187, II-188, II-189, II-190, II-191
Addictive	II-195
address	II-75, II-77, II-138, II-141, II-219, II-220, II-222, II-223, II-225
air	II-91, II-123, II-127, II-130, II-133, II-134, II-135, II-169, II-221
air cylinders	II-169
alternating	II-19
alarm system	I-202
Allegro Microsystems	I-90, I-92
Amp	II-24, II-26, II-31, II-56, II-143, II-167, II-235
Amplified	II-90, II-91
analog	I-219, I-220, I-231, I-234, II-115, II-123, II-133, II-145, II-187, II-190, II-195, II-196, II-197
AND gate	I-123, I-125, I-126
Astable	II-128
Audible	II-87, II-93
Audio	II-89, II-91, II-94, II-171, II-172, II-193, II-235
Automotive	II-154

B

Bar graph	II-195, II-235
base	I-42, I-71, I-146, II-149
Basic	I-17, I-18, I-60, I-123, II-230, II-233
BASIC	I-16, I-17, I-18, I-19, I-21, I-26, I-27, I-28, I-34, I-35, I-181, II-1, II-2, II-13, II-15, II-93, II-177, II-232, II-234
BASIC Stamp	I-16, I-17, I-18, I-19, I-20, I-21, I-25, I-26, I-27, I-28, I-29, I-30, I-31, I-33, I-34, I-35, I-37, I-38, I-39, I-40, I-41, I-45, I-60, I-68, I-69, I-75, I-79, I-80, I-84, I-129, I-148, I-149, I-181, I-187, I-196, I-233, II-1, II-2, II-13, II-15, II-93, II-177, II-232, II-234
BASIC Stamp II	I-17, I-18, II-1, II-2
BASIC Stamp Manual	I-21, I-26, I-28, I-181, II-232, II-234
Batteries	II-33
Battery	II-5, II-33
BCD	I-65

Index

bi-directional	I-120, II-216
binary	II-74, II-77, II-78, II-86, II-140, II-183
binary coded decimal	I-65
bi-polar	I-42, I-92, I-94, I-139
BJT	I-139, I-145, I-146, I-156
Blink	II-167, II-168, II-170
Blinker	II-167
Board of Education	I-29
Brain	II-175
break-before-make	I-55
BS-1	I-18
BS-2	I-18, I-29, I-41
buffer	I-115, I-119
building blocks	I-20
burglar alarm	II-219
Button	I-60

C

Carrier Board	I-26
CD22202	II-98
cellular phone	I-17
Centigrade	I-101
clip	I-29
clock	I-222, II-58, II-79, II-82, II-84, II-111, II-124, II-145
CMOS	I-115
Collector	I-146
Comparator	II-115, II-117, II-118
COM port	I-28
computer	I-15, I-17, I-25, I-28, I-89, II-9, II-229
connectivity	I-188
control	I-167, II-14, II-24, II-38, II-45, II-46, II-47, II-49, II-50, II-51, II-54, II-56, II-61, II-62, II-63, II-66, II-69, II-70, II-88, II-96, II-121, II-153, II-154, II-155, II-156, II-157, II-158, II-159, II-160, II-161, II-164, II-167, II-168, II-171, II-172, II-173, II-180, II-183, II-185, II-199, II-213, II-219, II-231
CPU	I-27
current	I-41, I-141, II-19, II-21, II-24, II-25, II-26, II-30, II-31, II-34, II-35, II-36, II-37, II-40, II-51, II-54, II-57, II-58, II-59, II-66, II-76, II-143, II-144, II-145, II-155, II-160, II-161, II-167, II-168, II-183
current-limiting	I-56, I-181

Index

D

DAC	I-231, I-234
data logger	I-21, II-52
DB-9	I-28, I-29
DC motors	I-161, I-164, II-153
de-bounce	I-60, I-61
decoder	II-96, II-99, II-101, II-102, II-222, II-223, II-225
development system	I-18, I-25, I-26, I-28
diagram	I-21, I-28
digital pot	I-185, I-188, I-191, II-54, II-163, II-167, II-168, II-171, II-199
DIP switch	I-16, I-29, I-30, I-65, I-116
Dipswitch	II-74, II-80, II-85, II-219, II-223, II-225, II-234
disk drives	I-89
distant planet	I-18
DPDT	I-55
DPST	I-54
DS1804	I-170, I-187, I-188, II-54, II-56, II-89, II-119, II-120, II-161, II-163, II-165, II-167, II-171, II-172, II-173, II-198, II-201
DS96174N	II-215, II-216, II-217
DTMF	II-6, II-93, II-94, II-95, II-96, II-97, II-98, II-99, II-101
D-type flip-flop	I-210
dynamic microphone	I-95

E

EDE1200	II-183, II-235
Edison	II-9
EEPROM	I-27, I-28
electric eye	I-71
electro-magnetic	I-155, II-143
electronic brain	I-19
emitter	I-42, I-146
encoder	II-219, II-225
ESD	I-37
Events	II-139, II-140, II-141

F

Fahrenheit	I-101

Index

fail-safe	II-63, II-64, II-155, II-175
field strength	I-93
filter capacitors	I-34
fish	II-123
flip-flop	I-210
flow	II-20, II-37, II-133, II-134, II-135, II-143, II-144, II-145
Ford	II-9
Frequency	II-5, II-87, II-94
full wave bridge	II-21, II-22
full-scale	II-190, II-204

G

garage door openers	II-219
Gate	I-123, I-146
ground	I-32, I-166, II-32, II-37, II-46, II-49, II-50, II-61, II-64, II-65, II-66, II-68, II-70, II-78, II-111, II-148, II-154, II-173, II-201, II-209, II-213, II-225

H

Hall-effect	I-89, I-90, I-91, I-92, I-93, I-94, II-58, II-108, II-143, II-144, II-145
hardware	I-15, I-16, I-17, I-18, I-19, I-20, I-23, I-28, I-132, I-170, I-176
H-bridge	I-164, I-165, II-163, II-164
heat sink	II-25, II-160
heat sinking	I-150, I-160
high impedance	I-118, I-119
HIH-3605-A	I-99
HT640	II-219, II-235
humans	I-181
humidity	I-99
hydraulic	II-127, II-130

I

I/O	I-16, I-21, I-31, I-33, I-37, I-38, I-39, I-40, I-41, I-42, I-43, I-45, I-46, I-47, I-49, I-55, I-56, I-58, I-63, I-79, I-85, I-110, I-111, I-115, I-116, I-118, I-119, I-120, I-129, I-155, I-168, I-170, I-174, I-181, I-195, I-196, I-198, I-200, I-206, I-207, I-208, I-209, I-210, I-211, I-212, I-213, I-214, I-215, I-216, I-217, I-231, I-233,

Index

	II-46, II-47, II-48, II-49, II-66, II-68, II-73, II-74, II-75, II-76, II-79, II-80, II-83, II-84, II-117, II-137, II-139, II-143, II-144, II-164, II-168, II-188, II-217, II-224
ignition systems	I-89
incandescent	I-135
incandescent lamps	I-135
inductive	II-63, II-66, II-153
inductors	II-40
Industrial control	I-16
infra-red	I-129, II-219, II-221, II-223
infrared LED	I-115, I-131
Input	I-51, I-105, I-195, II-5, II-6, II-71, II-137
input devices	I-20, I-57, I-187
input/output	I-21
instruction set	I-21
integrated circuit	I-27, I-34, II-24, II-185, II-193
interfacing techniques	I-17, I-19
intrusion detection	I-95
inventor	II-9, II-10, II-232
inverter	II-46, II-63, II-66, II-69, II-156, II-158
inverting buffer	I-43, I-116, I-117, I-140, I-150
IR	I-73
IRF511	I-131, I-147, I-148, I-150, II-61, II-63
isolated power	I-31, I-33

L

Lamp	I-16, I-135, I-137, I-138
Lazy	II-219
learning curve	I-26
LED	I-16, I-31, I-40, I-41, I-43, I-73, I-105, I-106, I-115, I-116, I-118, I-120, I-121, I-123, I-124, I-125, I-126, I-127, I-129, I-130, I-131, I-132, I-133, I-140, I-141, I-142, I-153, I-203, I-208, II-15, II-50, II-63, II-106, II-107, II-135, II-148, II-177, II-178, II-179, II-195, II-196, II-198, II-200, II-202
level shifting	I-50
light	II-9, II-50, II-63, II-105, II-110, II-134, II-144, II-170
light detection	I-76, I-77
light intensity	I-69

Index

linear	I-101, II-58, II-127, II-128, II-129, II-130, II-135, II-145
LM324	I-76, I-95, I-234, II-98, II-115, II-117
LM339	I-49, I-63, I-64, I-101
LM34	I-101
LM386	II-171, II-172
LM3909	I-132
LM3914N	II-195, II-201, II-202, II-203, II-205, II-235
LM555	I-60, I-61, I-80, I-136, I-154, I-170, I-184, II-88, II-90, II-128, II-159, II-160, II-161, II-163, II-164, II-165, II-167, II-168, II-169, II-170, II-175, II-176, II-177, II-178
LM7805	I-34, II-69
logic-level	I-60, I-148
low cost	I-18, I-164, I-167, II-168

M

machine controller	I-17, I-21
machine tools	I-89
magnetic field	I-89, I-90, I-92, I-95, I-155, I-161, I-177, II-58, II-144
make-before-break	I-55
MAX232CPE	II-210, II-235
MAX233CPP	II-211, II-235
MAX485	II-213, II-214, II-215
MAX666	II-51, II-52, II-53
Maxim	II-40, II-210, II-215
MC1488	II-209
mechanical relays	I-155
memory	I-27
Microchip Technology	I-27
Microcontroller	I-17, I-20, I-25, I-27, I-183, II-1, II-2, II-13, II- 15, II-16, II-46, II-231, II-233
Milky Way	I-16
Moisture	II-111, II-112, II-114
Momentary	I-54
Monitor	I-15, I-21, I-25, I-61, I-125, I-184, I-220
MOS	I-37, I-155
MOSFET	I-43, I-131, I-135, I-145, I-146, I-147, I-148, I-150, I-151, I-152, I- 156, I-164, I-182, I-183, II-46, II-47, II-50, II-57, II-58, II-59, II-61, II-63, II-111, II-113, II-153, II-154, II-155, II-157, II-160, II-163, II-167, II-173, II-183

Index

motor	I-161, I-173
Mouser Electronics	I-65, I-111, II-229
multi-tasking	I-132
mute	II-172

N

N.C.	I-53
N.O.	I-53
NAND gate	I-123, II-118
National Electric Code	I-175
National Semiconductor	I-101
N-channel	I-146, I-151, I-164, II-61, II-154, II-167, II-173
NEC	I-105
new products	I-18
noise immunity	I-93
noise threshold	I-58
NOR	II-99, II-101
normally-closed	I-53, I-58, I-59, II-36, II-86
normally-open	I-53, II-38
NPN	I-42, I-43, I-71, I-125, I-139, I-140, I-141, I-142, I-143, I-156, I-177, II-168

O

one-shot	I-61, II-178
on-state resistance	I-145, I-164, II-47, II-173
open collector	I-64, I-156
open-collector	I-111, I-117, I-142, I-143, I-150
operational amplifier	I-76, I-191
optical	II-106, II-109, II-129, II-130, II-154, II-230
optoisolator	I-105, I-106, I-107, I-153, I-166, II-50, II-63, II-66, II-67, II-70, II-148, II-155
output	I-113, I-153, I-203, II-19, II-20, II-21, II-22, II-23, II-24, II-25, II-26, II-30, II-31, II-34, II-35, II-40, II-41, II-42, 46, II-47, II-49, II-51, II-53, II-54, II-55, II-56, II-64, II-66, II-67, II-78, II-89, II-90, II-93, II-94, II-115, II-117, II-118, II-123, II-144, II-145, II-153, II-156, II-167, II-171, II-175, II-176, II-177, II-178, II-180, II-181, II-187, II-188, II-189, II-190, II-191, II-195, II-196, II-210, II-212, II-213, II-214, II-215, II-216, II-217, II-221
output devices	I-18, I-20

Index

P

Parallax	I-16, I-20, I-21, I-26, I-27, I-29, II-2, II-4, II-13, II-15, II-16, II-232, II-233
Patent	II-10
Pause	I-208
PBASIC	I-17, I-19, I-21, I-25, I-27
PBASIC interpreter	I-27
PCB	I-29, I-30
P-channel	I-146, I-151, I-152, II-46, II-58, II-61, II-113, II-153, II-154, II-155, II-160
perf-board	I-29
personal computer	I-18, I-26, I-28, I-29, II-183
photocell	I-67, I-68, I-69, I-79
photo-resistor	I-67
Phototransistor	I-71
PIC	I-27, II-233
plug-n-play	I-25
PNP	I-139, I-141, I-142, II-156, II-157
positive logic	I-159, II-46
positive supply	I-32, I-130, I-141
Pot	II-55, II-56, II-89, II-98, II-120, II-128, II-129, I-159, II-161, II-163, II-165, II-171, II-198, II-199, II-201, II-235
potentiometer	I-83, I-187, II-31, II-54, II-88, II-115, II-121, II-127, II-128, II-173, II-197
power supply	I-31, I-32, I-33, I-90, I-139, I-167, I-234, II-25, II-29, II-31, II-32, II-61, II-159, II-164, II-191
power-up reset circuit	I-28
pre-settable	II-161, II-163, II-167
printed circuit board	I-27, I-29
program	I-15, I-117, II-14, II-37, II-45, II-54, II-86, II-101, II-117, II-121, II-138, II-144, II-160, II-161, II-162, II-163, II-165, II-172, II-175, II-177, II-178, II-183, II-185, II-197, II-199
program samples	I-19
programming	I-19, I-26, I-28, II-13, II-138
programming cable	I-26, I-29, I-30
PS2501	I-105, I-106, I-153, II-234
pull-up	I-39, I-58, I-111, I-148, I-149, I-150, II-37, II-78, II-225
pull-up resistor	I-39, I-111, I-149, II-37, II-78, II-225
pulse	I-162, II-58, II-79, II-82, II-84, II-88, II-89, II-90, II-91, II-108, II-124, II-129, II-141,

Index

	II-145, II-159, II-161, II-172, II-173, II-175, II-177, II-178, II-181
Pulse Width Modulation	I-162
push-button	I-54, II-194
push-switch	II-78, II-80, II-81, II-86
push-wheel	II-77
PWM	I-162, I-163, I-164, II-196

R

R/C	II-195
radio-controlled	I-167
Radio Shack	II-113, II-230
RAM	I-27
ratio-metric	I-90
RC network	I-68
RCtime	I-68, II-128
real world	I-15, I-16, I-17, I-20, I-161, I-219, II-231
receiver	II-209, II-210, II-216, II-217, II-219, II-223
record	II-193, II-194
rectifier diode	I-129, II-21, II-23, II-26, II-35
remote data logging	I-75
resolution	II-106, II-109, II-188
re-start	II-175, II-177
re-triggerable	II-178
re-programmed	I-28
RH	I-99
Robotics	II-51
ROM	I-27, II-230

S

saw tooth	II-21
schematic diagram	I-19
Schmitt trigger	I-92, I-93, I-106, II-90, II-149
screen programming	I-21
security systems	I-89
semiconductor	I-71, I-145
serial	I-233, II-15, II-31, II-32, II-47, II-64, II-67, II-80, II-145, II-188, II-201, II-203, II-209, II-212, II-214, II-215
servo	I-167
shift register	I-208, I-209, I-210, I-211, I-213, I-214, I-216, I-217, II-47, II-64

Index

short circuit	I-65
shorting block	I-16
signal	II-51, II-99, II-118, II-155, II-175, II-177, II-209, II-210, II-213
software	I-15, I-16, I-18, I-20, I-26, I-29, I-60, I-195
solar	I-75, II-39, II-40, II-41, II-42
solar cell	I-75, I-76, I-77, I-78
soldering iron	I-22, I-37
solenoid	I-177
Solid State Relays	I-159
solid-state	I-137, I-170, I-173, II-88, II-89, II-121, II-130, II-131, II-161, II-193, II-197
source	I-32, I-33, I-40, I-75, I-105, I-106, I-109, I-111, I-129, I-131, I-141, I-142, I-146, I-151, I-152, I-173
Sonar	II-46
span	I-223, I-225, II-190
SPDT	I-54, I-55
speaker	I-181, II-87, II-91, II-173
speed controls	I-89
spikes	I-32, I-58, I-105, I-146, II-153, II-154
sprinkler systems	I-179
SPST	I-54
SSR	I-137, I-159, I-160, I-174, I-179, I-208
Stamp	I-16, I-17, I-18, I-19, I-20, I-21, I-25, I-26, I-27, I-28, I-29, I-30, I-31, I-33, I-34, I-35, I-37, I-38, I-39, I-40, I-41, I-45, I-60, I-68, I-69, I-75, I-79, I-80, I-84, I-129, I-148, I-149, I-181, I-187, I-196, I-233, II-13, II-15, II-16, II-34, II-37, II-51, II-54, II-75, II-89, II-91, II-93, II-128, II-177, II-195, II-213, II-229, II-232, II-233, II-234
Stamp Carrier Board	I-19
Stamp II	I-18, I-19, I-25, I-27, I-28, I-30, I-31, I-34, I-35, I-37, I-60
Stamps in Class	I-20
Stampworks	II-15, II-232
static	I-37
Stepper	II-183, II-184, II-185, II-235
successive approximation	I-221
switch bounce	I-59, I-60

Index

T
temperature	I-101
thermistor	II-79
timing cycle	I-60, II-159, II-161, II-167
transformers	II-19
transistor	I-139, I-145, II-9, II-41, II-47, II-50, II-63, II-66, II-69, II-156, II-157, II-160, II-168, II-181
Transmitter	II-209, II-215, II-216, II-219, II-220, II-221, II-225
tri-state	I-118, I-119
TTL	I-34, I-115, I-149, I-150

U
UGN3132UA	I-94
Ultrasonic	II-87
uni-polar	I-92, II-183

V
video game	I-21
voltage	I-64, I-65, I-102, II-19, II-20, II-21, II-23, II-24, II-26, II-30, II-31, II-32, II-33, II-34, II-35, II-36, II-39, II-40, II-45, II-47, II-50, II-51, II-53, II-54, II-55, II-56, II-57, II-62, II-63, II-70, II-78, II-90, II-115, II-116, II-117, II-118, II-120, II-123, II-145, II-147, II-153, II-154, II-158, II-167, II-168, II-188, II-190, II-195, II-196, II-197, II-198, II-199, II-201, II-209, II-210, II-211, II-212, II-213, II-235
voltage divider	I-46, I-67, I-68, I-81, I-85, I-86, I-224, I-225, I-228
voltage drop	I-33, I-135, I-136, II-47
voltage fluctuations	I-31, II-23
voltage regulator	I-28, I-34, II-23, II-24, II-30, II-34, II-40, II-54
voltage-controlled	I-43
volume controls	I-84

W
wall transformer	I-85, I-109, I-110, II-19, II-20, II-21, II-26, II-27, II-29, II-147, II-148

Index

wall transformers	I-109
wall wart	I-109
watchdog	II-175, II-176, II-179, II-180, II-181
water	II-111, II-114, II-115, II-117, II-133, II-134, II-135
Whitney	II-9
wiring diagrams	I-26
Wiring Simplified	I-175, II-233
word processor	I-21
wrist strap	I-37
write a program	I-17
Wozniak	II-9

Z

zener diode	I-46, II-35
zero	I-227, I-228, II-20, II-55, II-56, II-89, II-120, II-123, II-124, II-128, II-140, II-141, II-142, II-145, II-147, II-149, II-163, II-165, II-188, II-197, II-198, II-199, II-201
zero-cross	II-149
zero offset	I-102, I-227, I-229

Index